Go figure!

JUNE HAIGHTON

Lecturer in Mathematics
Gloscat – Gloucestershire College of Arts and Technology
Examiner for Application of Number

MARK ROWLAND

Lecturer in Mathematics
Richmond upon Thames College

Published in 2002 by:
Nelson Thornes Ltd
Delta Place
27 Bath Road
CHELTENHAM
GL53 7TH
United Kingdom

02 03 04 05 06 / 10 9 8 7 6 5 4 3 2 1

A catalogue record of this book is available from the British Library

ISBN 0 7487 6193 4

Illustrations by Bill Piggins
Page make-up by Tech Set

Printed and bound in Italy by Canale

Contents

Oh No... Not More Maths!

1.1 What is Application of Number?

'Oh no... Not more Maths!' Is this the thought uppermost in your mind as you start this course? Hopefully not, but if it is, take a look at the name of the course again: 'Application of Number'. The focus of the course is on the use of numerical skills that are **essential for everyday life**. In fact Application of Number is a Key Skill; in other words it is a skill that everyone needs, no matter what their job or interests, to cope with the numbers they will meet in life.

Think for a moment about everyday things that you do. You are probably using a variety of number skills already, though you may not think of them in this way. Even a simple activity such as planning a journey involves working with numbers. You may need to work out how long the journey will take and what time you should set off. If you are going by bus or train you might need to consult timetables. Finding the cheapest way to reach your destination may require several calculations. Cooking a meal involves measuring or estimating quantities and shopping involves money calculations. Even understanding the news on television often requires some knowledge of statistics. In all of these activities you are applying number skills.

Your hobbies or interest may also involve the 'Application of Number'. For example, if you are interested in a sport such as football, you will no doubt study the results and league tables. The internet provides an almost unlimited supply of information about football and other sports and hobbies. Selecting, organising and interpreting the available data is an important skill. Making sense of such data will be a feature of your course. In fact it is possible to use it as the basis for some of the work you do for your **portfolio**. Building a portfolio is discussed in Chapter 4.

What sort of career do you wish to follow? Most jobs include the use of numbers in one form or another. In the construction industry, bricklayers, surveyors and architects need to measure accurately and make sensible approximations. In the financial world, bank clerks insurance agents and accountants need to work with money. In the leisure industry, managers of sports centres, cinema complexes and other facilities need to set targets and budgets. Journalists, politicians and managers in every industry need to be able to understand and use statistical information. The list is endless! Even jobs that appear to have nothing to do with numbers, such as peeling potatoes in a kitchen will

have a payslip at the end of the week to be checked. Performance in many jobs is measured in some way using numbers or percentages. The person peeling potatoes might be assessed by measuring how long it takes to peel a sack of potatoes and calculating the percentage of potatoes discarded in the peelings. If you have a job, you should understand the numerical measures that are used to assess your work as well as the calculations on which your wage packet is based. If you work overtime you should check that you have been paid the correct amount.

The ability to **apply** numerical techniques is a valuable asset in many areas of life. So you can see that this course is indeed NOT just more Maths – it's the **Application** of Number!

1.2 Why you need this book

This book will help you to develop your number skills in a way that is relevant to everyday life. Probably your main aim in using the book is to complete the Application of Number unit successfully. Each of the three main Key Skill units – Application of Number, Communication and Information Technology – can be taken at different levels. The Application of Number work in this book is focussed on levels 2 and 3. At both of these levels the unit is assessed by means of an external test (Part A) and a portfolio of evidence (Part B). The specifications for both level 2 and level 3 can be found at the back of the book.

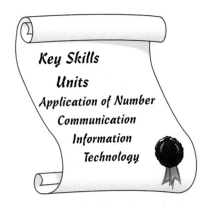

The test paper is the same whichever Awarding Body (e.g. AQA, Edexcel or OCR) your school or college is registered with. At level 2, the test is in the form of forty multiple-choice questions. At level 3, each problem is set in a real life context and is broken down into a series of questions. Calculators are allowed in the level 3 test but not in the level 2 test.

The portfolio of evidence consists of work that you complete during the course. This work will consist of one or more substantial activities in which you obtain and interpret information, use the information in calculations and explain how the calculations meet the purpose of the activity. There are a wide variety of possible activities that could be used. Ideally, those that you do should be chosen because they are relevant to your other studies or interests. At the end of the course your portfolio will be assessed by your teacher, and might then be checked by the Awarding Body.

This book explains and gives practice in all the mathematical methods needed for Application of Number at levels 2 and 3. The book includes helpful hints on how to prepare for and pass the test. It also gives advice on how to produce a good portfolio and suggests contexts in which you can base your portfolio work. Whether your main aim is to simply improve your number skills or to pass the Application of Number unit, this book will help you to achieve it.

1.3 How to use this book

If you are studying for the Application of Number unit, work through Chapters 2 and 3 to ensure that you study and practise all the techniques that will be included in the test. Most sections are needed for both level 2 and level 3. Sections that are required for level 3, but not for level 2 are indicated by this icon:

 LEVEL **3**

If you are working towards level 2, you should in general do calculations **without the aid of a calculator**. This will help you to develop your number skills and prepare you for the test. Questions with the level 3 icon are more complex, usually involving more difficult arithmetic. If you are preparing for level 3, use a calculator for these questions if you wish.

Chapter 2 begins with the techniques used in calculations. It then explains how to use equations to solve problems before concluding with some topics in shape and space. Chapter 3 deals with statistical analysis and interpretation. An overview of statistical charts and graphs is followed by details of how to organise, present and interpret statistical information.

The order of the sections in each chapter has been planned carefully so that you have the knowledge at each stage to cope with the next. Occasionally you may find that you need extra help with some of the basic methods used. Tricks of the Trade (ToTT) gives a summary of all of the numerical techniques you are expected to be familiar with at the start of your course. These include non-calculator methods involving whole numbers, fractions, decimals, ratios, percentages and negative numbers (ToTT.1–6). At points in the text where these techniques are needed the relevant sections of Tricks of the Trade are identified. If you are unsure of these methods you can study them as the need arises.

During the course you will need to produce evidence for your portfolio. Consult Chapter 4 before you start any portfolio work. It gives helpful advice about how to organize your work and how to obtain useful data. It also gives many ideas for activities that you can use to produce the sort of evidence you need. When working on your portfolio you can use a calculator wherever necessary at both levels 2 and 3. Tricks of the Trade Section 7 (ToTT.7) explains all the calculator methods you will need. Refer to this section whenever you need help.

Chapter 5 gives useful information about the external test. Look at this chapter well before the date of the test so that you have time to put the advice on revision into practice.

A summary of the structure of the book is given in the table below:

For guidance on:	Use chapters:
the external test (Part A)	2, 3, 5, ToTT
your portfolio (Part B)	2, 3, 4, ToTT
Application of Number specifications	Specification guides

Throughout the book the margin at the right-hand side of the page is used to give extra helpful notes. These may be hints on how to proceed with a problem or notes on how to check calculations. They may also be references to other sections of the book containing related topics. It is vital that you get into the habit of checking your work. There are various ways of doing this. For example, you could work out the answer using an alternative method, carry out an approximate calculation or do an inverse check. These methods are explained more fully in Tricks of the Trade Section 8 (ToTT.8) and are used throughout the book. ToTT.8 also gives advice on accuracy and shows how to round numbers using decimal places or significant figures.

Important facts that you should learn are accompanied by this icon:
This will make it easy for you to revise what you need to know before the test.

The Nelson Thornes website provides useful supplementary material such as practice test questions, useful website addresses and sets of data in spreadsheet format that can be used for practising statistical techniques. The address is www.nelsonthornes.com/gofigure.

This book, together with the Nelson Thornes website and your hard work, will provide you with everything you need to be successful in Application of Number. Good luck!

2.1

Calculation Techniques

2.1.1 Working with fractions

'of' means multiply

Example

A pair of trainers normally cost £60.
They are advertised in a sale at $\frac{2}{3}$ their normal price.
Find the sale price.

Here's how...

Sale price $= \frac{2}{3} \times £60$

Divide 60 by 3 to find $\frac{1}{3}$: $60 \div 3 = 20$

Multiply by 2 to find $\frac{2}{3}$: $2 \times 20 = 40$

Answer: Sale price $= £40$

> **hint**
> For fraction methods see Tricks of the Trade 2&7.

> **hint**
> Always show your working.

> **checkpoint**
> or... multiply by 2 then divide by 3:
> $60 \times 2 = 120$
> $120 \div 3 = 40$

'of' means 'multiply'
amount = fraction × total

Now try these...

1   Find the following amounts.
    a $\frac{3}{4}$ of £8   **b** $\frac{2}{5}$ of £20   **c** $\frac{2}{3}$ of £210   **d** $\frac{5}{12}$ of £36   **e** $\frac{3}{4}$ of £6
    f $\frac{5}{6}$ of £90   **g** $\frac{3}{8}$ of £20   **h** $\frac{1}{4}$ of £9   **i** $\frac{3}{5}$ of £7   **j** $\frac{2}{3}$ of £20 (to nearest pence)

2   In a school of 960 pupils, two-thirds study a foreign language.
    Find how many pupils in the school study the foreign language.

3   The population of the UK in 1972 was approximately 56 million.
    Six in every seven people lived in England. Find the number of people living in England in 1972.

4   How many minutes is seven-twelfths of an hour?

5   A lottery winner scoops £840 000. She spends three-eighths of the winnings to buy a new house.
    a How much does the new house cost?
    b How much of the winnings does she have left after buying the new house?

6   Two in every fifteen passengers travel on public transport without a valid ticket.
    A train holds 461 people including eleven ticket inspectors and other members of staff.
    Estimate the number of passengers on the train who are travelling without a valid ticket.

Calculations may involve several steps

For example, before the required amount can be found, it may be necessary to first find the total…

Example

Three quarters of the fish in a garden pond suffer from fin rot.
The pond contains 5 goldfish and 7 carp.
Find the number of fish with fin rot. Show all the steps in your working.

Here's how…

Total number of fish in the pond $= 5 + 7 = 12$
No. of fish with fin rot $\qquad = \frac{3}{4} \times 12$
$\qquad\qquad\qquad\qquad\quad = 9$

hint

$12 \div 4 = 3$
$3 \times 3 = 9$

Answer: 9 fish have fin rot

…or to find all the other parts first.

Example

A kitchen drawer stores 30 pieces of cutlery. $\frac{1}{3}$ are knives and $\frac{2}{5}$ are forks.
The rest of the cutlery are spoons.
How many spoons are in the drawer?

Here's how…

First find the number of knives and forks:
No. knives $\qquad = \frac{1}{3} \times 30$
$\qquad\qquad\quad = 10$
No. forks $\qquad\; = \frac{2}{5} \times 30$
$\qquad\qquad\quad = 12$
No. knives + forks $= 10 + 12$
$\qquad\qquad\quad = 22$
No. spoons $\qquad = 30 - 22$
$\qquad\qquad\quad = 8$

checkpoint

Makes sense:
22 is less than 30
(= total)

checkpoint

Check: $\frac{1}{3} + \frac{2}{5} = \frac{5}{15} + \frac{6}{15} = \frac{11}{15}$
So the other $\frac{4}{15}$ are
spoons. $\frac{4}{15} \times 30 = 8$

Answer: 8 spoons

Now try these…

1 Express the following as single fractions:

   a $\frac{1}{2} + \frac{2}{5}$ **b** $\frac{2}{7} + \frac{1}{4}$ **c** $\frac{3}{5} - \frac{2}{9}$ **d** $\frac{7}{8} - \frac{3}{4}$

2 Two fifths of a class of students wear glasses. There are 14 boys and 11 girls in the group. How many students in the group wear glasses?

3 Two ninths of a delivery of apples supplied to a supermarket are damaged in transport. The apples are supplied in 5 crates of 18. How many apples are damaged?

4 Of the 48 people registered to take their driving test one day, $\frac{3}{4}$ passed and $\frac{1}{6}$ failed. The remainder were unable to attend the test. How many people did not attend the test?

5 Weather records for November 2000 described each day's weather as either Rain, Snow, or Dry. $\frac{2}{3}$ of November was classed as Rain, and $\frac{3}{10}$ as Dry. On how many days in November did it snow?

You may need to find a fraction of a fraction…

Example

A committee has 40 people. $\frac{3}{5}$ have brown hair, of which $\frac{3}{4}$ have blue eyes.
Find the number of people with brown hair and blue eyes.

Here's how…

$$\text{Number of people with brown hair} = \frac{3}{5} \times 40$$
$$= 24$$

$$\text{Number of people with brown hair and blue eyes} = \frac{3}{4} \times 24$$
$$= 18$$

Answer: 18 people

checkpoint

or… multiply the given fractions:
$\frac{3}{4} \times \frac{3}{5} = \frac{9}{20}$
$\frac{9}{20} \times 40 = 18$

…or to divide an amount by a fraction…

hint

To divide by a fraction, invert and multiply.
$\frac{1}{2}$ inverted $= \frac{2}{1} = 2$

Example

An empty three litre container is to be filled with water using a half litre jug.
How many times must the jug be used in order to fill the container?

Here's how…

The answer is the number of times $\frac{1}{2}$ goes into 3.

$$\text{No. times jug is used} = 3 \div \frac{1}{2}$$
$$= 3 \times 2$$
$$= 6$$

Answer: The jug is used 6 times

checkpoint

makes sense:
$6 \times \frac{1}{2} = 3$

Now try these…

1 Express the following as a single fraction:
   a $\frac{3}{5} \times \frac{2}{7}$ **b** $\frac{2}{9} \times \frac{1}{4}$ **c** $\frac{4}{5} \times \frac{9}{10}$

2 Work out the following:
   a $6 \div \frac{1}{3}$ **b** $8 \div \frac{4}{5}$ **c** $15 \div \frac{3}{5}$

3 A herd has 45 cows. $\frac{1}{3}$ of the cows are Friesians, of which $\frac{4}{5}$ are ready for milking.
   a How many Friesians are there?
   b How many Friesian cows are ready for milking?

4 A group has 13 boys and 17 girls. $\frac{2}{3}$ of the group pass GCSE Maths, of which $\frac{2}{5}$ achieve grade A*.
   How many pupils in the group achieve a GCSE grade A* in Maths?

5 A town has a population of 150 000. $\frac{1}{5}$ of the town are not eligible to vote. In an election, $\frac{3}{5}$ of those eligible to vote do so. $\frac{1}{4}$ of all votes cast are for the *Purple Party*.
   a How many people in the town vote?
   b How many votes does the *Purple Party* receive?

6 A clock chimes every quarter of an hour, starting at 12 noon. How many times does the clock chime between 3:20 pm and 8:50 pm on the same day?

7 A $2\frac{1}{4}$ litre bucket was used to fill a children's paddling pool. The pool holds just over 20 litres.
   How many times was the bucket filled and emptied into the pool?

LEVEL 3

8 A fence 14 m long consists of a series of fence posts and horizontal planks. Each pair of fence posts are $\frac{7}{8}$ m apart and are joined by three horizontal planks.
   a How many fence posts are used to build the fence?
   b How many horizontal planks are used in total?

9 An entrance exam to a law firm consists of two tests. A group of 280 students sit both tests. They must pass both tests to gain automatic entry to the firm. Some pass and fail fractions are shown in Table 2.1:
   a How many students gained entry to the firm?
      Altogether $\frac{3}{4}$ of the students passed Exam 1.
   b Copy and complete the table, giving fractions in their lowest terms.
      Students who pass Exam 1 but fail Exam 2 are allowed a resit.
   c How many students are allowed to resit Exam 2?
      The resit students have a pass rate of $\frac{2}{3}$.
   d How many students in total gain entry to the firm?

		Exam 1	
		Pass	Fail
Exam 2	Pass	$\frac{3}{5}$	
	Fail		$\frac{1}{5}$

Table 2.1

An amount can be increased or decreased by a fraction

When increasing by a fraction, **add** the increase to the original amount.

Example

The value of a £3000 investment increases by a half.
Find the new value.

Here's how...

Increase $\quad = \frac{1}{2} \times £3000$
$\qquad\qquad = £1500$

Add the increase to the original amount
New value $\quad = £3000 + £1500$
$\qquad\qquad = £4500$

Answer: New value = £4500

checkpoint

or... find $1\frac{1}{2} \times 3000$
$= \frac{3}{2} \times 3000$
$= 4500$

checkpoint

Makes sense: answer is greater than £3000

A decrease must be **subtracted** from the original amount.

Example

A digital camera normally costs £120.
In a sale, the price is reduced by one-third. Find the sale price.

Here's how...

Decrease $\quad = \frac{1}{3} \times £120$
$\qquad\qquad = £40$

Subtract the decrease from the original amount
Sale price $\quad = £120 - £40$
$\qquad\qquad = £80$

Answer: Sale price = £80

checkpoint

or... find $\frac{2}{3} \times 120 = £80$

checkpoint

Makes sense: answer is less than £120

Now try these...

1 **a** Increase **i** £12 by $\frac{1}{3}$ **ii** £55 by $\frac{1}{5}$ **iii** £30 by $\frac{2}{3}$ **iv** £18 by $\frac{3}{4}$
   b Decrease **i** £28 by $\frac{1}{7}$ **ii** £90 by $\frac{2}{3}$ **iii** decrease £100 by $\frac{2}{3}$ then increase the answer by $\frac{2}{5}$.

2 Sue earns £180 per week. Her pay is increased by a third. What is her new weekly wage?

3 The flow of traffic on one of Britain's motorways in 1991 was 54 000 vehicles per day.
   By 1999, the daily flow had increased by $\frac{1}{4}$. Find the daily flow in 1999.
   Give your answer to the nearest thousand vehicles per day.

4 The UK infant mortality rate in 1963 was 24 deaths per thousand births. By 1993, medical advances
   reduced the infant mortality rate per thousand births by two thirds.
   Find the infant mortality rate per thousand births in 1993.

5 Shares in a pharmaceutical company were bought for £1600. In the first year, the shares increased
   in value by $\frac{5}{8}$. In the second year, their value decreased by $\frac{3}{4}$.
   How much were the shares worth after:
   a one year   **b** two years?

6 The test results of 20 students are recorded in Table 2.2.
   The test is marked out of 40. Half of the students have passed.
   a Find the pass mark for the test.
   b Mark the following as true or false:
      i Increasing the pass mark by $\frac{1}{5}$ will decrease the number of passes by $\frac{3}{5}$.
      ii Decreasing the pass mark by $\frac{1}{5}$ will increase the number of passes by $\frac{2}{5}$.

30	24	32	17	25
27	40	35	26	20
36	29	31	14	39
11	35	37	30	23

Table 2.2

LEVEL 3

7 Since 1997, the number of people diagnosed as having dyslexia has increased by $\frac{1}{4}$ each year.
   In 1997 there were 1.2 million cases.
   a Calculate the number of cases diagnosed in 1999, using an appropriate level of accuracy.
   b Is the only explanation for the increase that dyslexia is becoming more common?
   Justify your answer.

8 Some information about deaths in Archdale is shown in Table 2.3.
   Figures in brackets give the fractional change year on year.

Year	1998	1999		2000	
Road accidents deaths	...	...	$(+\frac{1}{5})$	...	$(+\frac{3}{8})$
Total number of deaths	600	...	$(-\frac{1}{10})$	...	$(-\frac{1}{9})$

Table 2.3

In 1998, 2 in every 15 deaths were caused by road accidents.
a Calculate the number of road accident deaths in 1998.
b Calculate all the other entries in the table.

An expert makes the following statement:

"The proportion of road accident deaths has more than doubled between 1998 and 2000."

c State whether you agree or disagree with the expert. Justify your answer.

2.1.2 Working with ratios and proportion

An amount can be divided in a given ratio

hint

For ratios see ToTT.4.

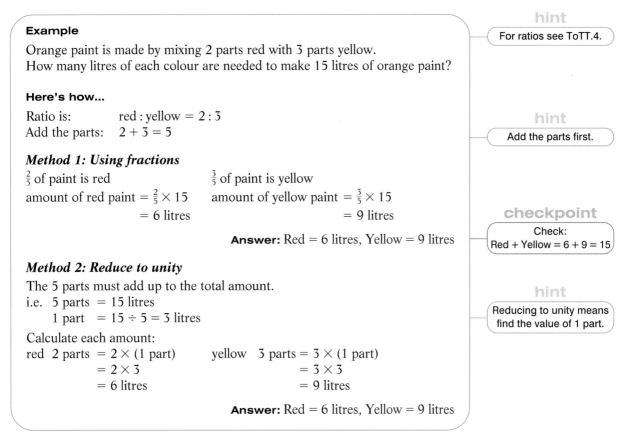

Example

Orange paint is made by mixing 2 parts red with 3 parts yellow.
How many litres of each colour are needed to make 15 litres of orange paint?

Here's how...

Ratio is: red : yellow = 2 : 3
Add the parts: 2 + 3 = 5

hint

Add the parts first.

Method 1: Using fractions

$\frac{2}{5}$ of paint is red $\frac{3}{5}$ of paint is yellow

amount of red paint = $\frac{2}{5} \times 15$ amount of yellow paint = $\frac{3}{5} \times 15$

            = 6 litres   = 9 litres

Answer: Red = 6 litres, Yellow = 9 litres

checkpoint

Check:
Red + Yellow = 6 + 9 = 15

Method 2: Reduce to unity

The 5 parts must add up to the total amount.
i.e. 5 parts = 15 litres
      1 part   = 15 ÷ 5 = 3 litres

Calculate each amount:
red 2 parts = 2 × (1 part)   yellow 3 parts = 3 × (1 part)
        = 2 × 3   = 3 × 3
        = 6 litres   = 9 litres

Answer: Red = 6 litres, Yellow = 9 litres

hint

Reducing to unity means find the value of 1 part.

NB the ratio 2:3 can also be written as 1:1.5 (dividing both parts by 2).

Now try these...

1 Divide the following amounts in the given ratio.
   a £40 in the ratio 2:3   **b** £160 in the ratio 3:5   **c** £2400 in the ratio 2:3:7

2 One day, Mr Reynolds receives personal and business calls on his mobile phone in the ratio 2:5.
   He receives a total of 21 calls.
   How many: **a** personal calls   **b** business calls   does he receive that day?

3 Each week, Mark spends £3 and Jane spends £5 on the *National Lottery*. They agree to share out all
   their winnings in the ratio of their contributions. One week one of their tickets wins £1680.
   How much should   **a** Mark receive   **b** Jane receive?

4 A shop stocks two types of juice, *orange* and *mango*. Out of 63 cartons, 18 are unsold.
   The ratio of *orange* to *mango* sales are 5:4.
   How many cartons of: **a** *orange juice*   **b** *mango juice*   are sold?

5 An absent minded doctor cannot remember how many patients are on his register. Records show the
   ratio of males to female patients is 2:3 and the ratio of private to NHS patients is 2:5.
   How many patients are on the register?
   Choose from: **A** 60   **B** 63   **C** 70

LEVEL 3

6 65 people are tested for colour blindness. Here are some of the results.

	Male	Female	Total
Colour blind	4	6	
Not colour blind			
Total			65

Table 2.4

a How many colour blind people are there altogether? Copy the table and enter this value.

For the non-colour blind people, the ratio of males to females is 5:6.

b Complete all parts of the table.

c Does the table suggest that colour blindness is more common in males than in females? Justify your answer.

7 Life expectancy in years (from birth) for people born in the UK is shown in Table 2.5.

Year	1982	1987	1992	1997
Male	68.9	71.1	73.2	74.5
Female	75.1	77.0	78.6	79.0

Table 2.5

For example, a girl born in 1992 can expect to live 78.6 years.

a Express the ratio of male to female life expectancy in the form 1:x for each year. Work to two decimal places.

b Make two comments about male and female life expectancy from 1982 to 1997.

Boys born in 2002 can expect to live 76.8 years.

c Estimate the life expectancy of a girl born in 2002, stating any assumptions made.

Reduce to unity to find one part

Example

The ratio of men to women at a conference is 3:4.
There are 18 men at the conference.
Find the number of: **a** women at the conference
b people at the conference.

Here's how...

a 3 parts are men.
            3 parts = 18 (men)
    [÷3]   1 part = 18 ÷ 3 = 6 people
Women are represented by 4 parts.
    [×4]   4 parts = 4 × 6 = 24 people

Answer: 24 women are at the conference

b Add the parts: 3 + 4 = 7. The total is represented by 7 parts.
        1 part = 6 (people)   {from part **a**}
        7 parts = 7 × 6 = 42 people

Answer: 42 people are at the conference

hint

men:women = 3:4

hint

Make use of previous answers.

checkpoint

Check:
Men + women = 18 + 24
= 42

Now try these...

1 An inheritance is divided between two children in the ratio 2:3. The smaller share is worth £8460.
   What is the value of: **a** the larger share   **b** the total inheritance?

2 Alice, Bob and Carol are playing Monopoly. Bob has £2500.
   The amount of money (in £) each person has is in the ratio Alice:Bob:Carol = 3:5:6
   a How much does Alice have?   **b** How much *more* does Carol have than Bob?

3 In the village of *Rumoursville* in 2000, the marriage to divorce ratio was 5:3.
   There were 120 marriages in 2000. How many *people* got divorced in 2000?

4 A hockey team plays each of 21 other teams twice during a season.
   For every 3 games the team wins, it loses 2. At the end of the season, the team has won 24 games.
   a How many games has the team lost?   **b** How many games has the team drawn?

5 The minimum age for joining a youth club
   is 14 years. Some results from a survey of all
   members are given in Table 2.6:

Age (in years)	14	15	16 or over
No. members	12	27	

Table 2.6

   The ratio of under-16 year olds to those
   16 or over is 3:5.
   a How many members are 16 or over?
   b How many members does the club have in total?

   The survey is repeated one year later. Nobody has joined or left the club in that time.
   c What is the ratio of under-16 year olds to those 16 or over?
   Choose from: **A** still 3:5   **B** 12:65   **C** 3:23

Reduce to unity to scale any quantity

Example

Two theatre tickets cost £15. Find the cost of three tickets:

Here's how...

Information given: 2 tickets = £15
Reduce to unity   : 1 ticket   = £15 ÷ 2 = £7.50
Scale up by 3   : 3 tickets = 3 × £7.50
                                       = £22.50

Answer: Cost = £22.50

hint
Reduce to unity.

checkpoint
Makes sense:
answer is more than £15

Jobs are often advertised *pro rata*. This means payment is proportional to the number of hours worked.

Example

A normal working week is 40 hours. A job is advertised at £16 000 pro-rata
per year. The position is filled by a part-timer who works 25 hours a week.
How much does he earn per year?

Here's how...

	working week (hours)	salary (£)
Information given:	40	16 000
Reduce to unity :	1	16 000 ÷ 40 = 400
Scale up by 25 :	25	25 × 400 = 10 000

Answer: Salary = £10 000 per year

checkpoint
or... Part-timer works
$\frac{25}{40} = \frac{5}{8}$ of the week.
$\frac{5}{8} \times 16\,000 = 10\,000$

Now try these...

Assume all quantities increase or decrease in proportion.

1 Three travelcards cost £9.60. What is the cost of 4 travelcards?

2 A microwave defrosts a 200 g piece of steak in 180 seconds.
  How long does the microwave take to defrost a 450 g piece of steak?

3 A normal 5 day working week is 40 hours less one hour a day for lunch (unpaid).
  A job is advertised at £21 000 pro rata per year. The position is taken by someone who works
  25 hours per week. How much does the part-timer receive per year?

4 An international phone card is used to call Japan. The call lasts 6 minutes and costs £1.50.
  What is the longest call to Japan that can be made with a £10 card?

5 Jane pays £12 per month in line rental for her mobile phone. A four minute peak rate call costs £1.60.
  Jane limits herself to spending £50 per month on her phone.
  If all calls made are at peak rate, how many minutes can she use her phone per month?

6 A 180 g box of 'Sud-U-Like' soap powder costs £1.50.
  a How many grams of 'Sud-U-Like' is this per £1?

  Its rival powder 'Keen To Clean' comes in 285 g boxes and is £1 more expensive.
  b Which brand offers the best value for money?

7 A shoe shop assistant claims shoe size increases in proportion to height. One customer takes
  size 8 shoes and is 170 cm tall. The next customer takes size 10 shoes.
  a If the assistant's claim is correct, how tall would the second customer have to be?
  b Do you think the assistant's claim is correct?

LEVEL 3   Several quantities can be scaled in the same proportion

Example

The information below shows the nutrition label on a 180 g can of tuna:

> **Contents per 100 g of product: Protein = 30 g Fat = 0.5 g**

How much protein and how much fat is in the can of tuna?

Here's how...

The values given are for 100 g of tuna. The can holds 180 g of tuna.

	tuna (g)	protein (g)	fat (g)
Information given :	100 g	30	0.5
Reduce to unity :	1 g	$30 \div 100 = 0.3$	$0.5 \div 100 = 0.005$
Scale up by 180 :	180 g	$180 \times 0.3 = 54$	$180 \times 0.005 = 0.9$

Answer: The can contains 54 g of protein and 0.9 g of fat

hint
> For decimal methods see ToTT.3.

checkpoint
> Makes sense: 180 g is nearly twice 100 g

Now try these...

Assume all quantities increase or decrease in proportion.

1 On average, a 500 word report contains 5 spelling and 4 grammatical errors.
  Find the number of expected errors of each type in a report with 8000 words.

2 The following ingredients are used to make 2 litres of non-alcoholic punch:
     1.3 litres of orange juice   0.7 litres of pineapple juice   4 oz of fruit
  Use the conversion 1 pint ≈ 0.6 litres to calculate the quantities of each ingredient needed for
  4 pints of punch. Give answers (in litres and ounces) using an appropriate degree of accuracy.

3 Sam has a recipe for spaghetti napoletana. The quantity of the main ingredients needed for four people are:

Ingredients   spaghetti (500 g)   tomatoes (300 g)   minced beef (240 g)
Preparation time = 30 minutes (not including cooking)

Sam prepares spaghetti napoletana for himself and five friends.
a Find the amount of each main ingredient needed.
b How long (in minutes) does it takes Sam to prepare the meal (not including cooking)?

4 Table 2.7 shows some of the nutritional information on a packet of biscuits:
a Use the information to find:
  i the total energy in the pack
  ii the total fat in the pack.
b Use your answers to part **a** to estimate the number of biscuits in the pack.

Pack weight = 450 g		
Nutritional information:	Per 100 g	Per biscuit (approx.)
Energy	2010 kJ	270 kJ
Fat	19.6 g	2.7 g

Table 2.7

2.1.3 Working with percentages

Finding a percentage of an amount may involve more than one step

Example

In a group of students, 58 are male and 62 are female.
20% of the group study physics. How many study physics?

Here's how...

Step 1: Find the total 58 + 62 = 120

Step 2: Calculate the 'amount'.
    No. studying physics = 20% of 120
                         = 24

Answer: 24 students study physics

hint

For percentage methods see ToTT.5.

hint

$\frac{2\emptyset}{1\emptyset\emptyset} \times 12\emptyset = 24$

checkpoint

10% of 120 = 12
20% of 120 = 24

Now try these...

1 A health club has 27 full-time members and 43 part-time members. 20% of members have medical insurance. How many members have medical insurance?

2 1200 people vote in an election. 40 ballot papers are spoiled and discarded. Jeremy Archer gets 15% of the votes counted. How many votes does Mr Archer get?

3 3500 home fans and 1500 away fans attend a football match. 40% of the crowd buy tickets at the gate. The rest are season ticket holders.
a How many people buy tickets at the gate?   **b** How many people are season ticket holders?

4 In a gymnastic competition, five judges each award marks out of 6 for each performer. The gold medalist receives 80% of the total possible score. Some of the judges' marks are given in the table.
Calculate the mark awarded by Judge 3.

Judge	1	2	3	4	5
Mark	3	5		6	6

Table 2.8

5 Sally's monthly take-home pay is £1104. Deductions are £207 for income tax and £109 for National Insurance. Sally wants to pay 5% of her gross salary (i.e. *before* deductions) into her pension fund. How much does Sally contribute to her pension fund per month?

Now try these...

6 A company employs 8 men and 12 women. 60% of the employees drink only decaffeinated coffee. Which of the following statements must be true? Explain why.
   i Over half the employees drink decaffeinated coffee.
   ii At least one man drinks decaffeinated coffee.
   iii 40% of the employees drink ordinary coffee.

7 Table 2.9 shows details of money dispensed by a cash machine:
   a How many notes are dispensed altogether?
   b What is the total value of money dispensed?
   $2\frac{1}{2}$% of the notes dispensed are counterfeit £50 notes.
   c What was the actual value of the money dispensed?:
      Choose from: **A** £1580   **B** £1638   **C** £1780

Note value	£10	£20	£50
Number of notes dispensed	34	32	14

Table 2.9

8 The suppliers of a pie filling claim their product contains '*no less than 80% meat*'. The filling consists of: 550 g meat, 82 g tomatoes, 38 g mushrooms, 25 g other
Are the suppliers telling the truth?

9 The table shows some of the medals won at an Olympic Games. Use the following information to complete the table.

30% of the gold medals in Table 2.10 were won by the USSR.

10% of the medals in the table are UK bronze medals.

15% of the USA's medals were silver.

	Gold	Silver	Bronze	Total
UK	1	4		
USA				
USSR				27
Total	40	19	21	

Table 2.10

An amount can be increased or decreased by a percentage

Amounts are often altered by a percentage. For example, a water company may increase its charges to customers by 8%. A shopkeeper may reduce prices by 25% in a sale. There is more than one method for increasing an amount by a percentage.

Example
Increase £50 by 10%.

Here's how...
Method 1: increase   = 10% of £50 = £5
    new amount = original + increase
               = £50 + £5
               = £55

                                Answer: New amount = £55

hint
Short cut: to find 10%, divide by 10

checkpoint
Makes sense: answer is more than £50

Method 2: new amount = original amount + increase
                    = 100% of £50 + 10% of £50
                    = 110% of £50
                    = 1.1 × £50
                    = £55

                                Answer: New amount = £55

checkpoint
Add the %:
100% + 10% = 110% = 1.1
1.1 × 50 = 55

hint
For decimal methods see ToTT.3.

Note: Method 2 is more direct but requires the use of decimals.

A reduction must be subtracted from the original amount.

Example

Mr Warren earns £20 000 per year. He pays 15% income tax.
How much does Mr Warren earn per year after paying tax?

Here's how...

Tax paid
$$= 15\% \text{ of } 20\,000$$
$$= £3000$$

Amount earned per year after tax
$$= £20\,000 - £3000$$
$$= £17\,000$$

Answer: Net income = £17 000

checkpoint

Short cut:
15% = 10% + 5%
Tax = £2000 + £1000
= £3000

checkpoint

or... subtract the %:
100% − 15% = 85% = 0.85
0.85 × 20 000 = £17 000

Now try these...

1 A pint of milk in a corner shop costs 40p. The shopkeeper increases its price by 5%.
   How much does the milk cost after the increase?

2 A bed normally costs £350. The bed is on sale in two different shops.
     Shop 1: sale price = 65% of £350   *Shop 2: sale price = 40% off £350*
   Without working out the sale prices, say which shop offers the best deal.

3 A photograph developing firm offers the following print sizes:
     Regular = 10 cm by 15 cm   Enlarged = 30% longer and 30% wider
   What size is an enlarged print?

4 The recommended retail price (r.r.p) for detergent is £3.20 per bottle. A local store sells the detergent
   at 10% above the r.r.p. A supermarket sells the same detergent at 5% below the r.r.p
   a What price does the local store charge for the detergent?
   b How much more expensive (in pence) is the detergent in the store compared to the supermarket?

5 In September, a college has 4800 students. 10% of the students leave college during the first term.
   A further 5% of students leave during the second term. No new students join the college.
   How many students are registered with the college after: **a** 1 term   **b** 2 terms?

6 Sunlight has an intensity of 20 000 candela (cd). In London during the 1999 eclipse, 95% of the sun
   was covered by the moon. The remaining sunlight was further reduced by 5% due to cloud cover.
   What was the light intensity (in cd) seen in London during the eclipse?

7 Mrs Warren earns £33 280 per year. She pays a total of 25% in income tax and National Insurance.
   a Find Mrs Warren's net annual income (after deductions).
       Mrs Warren also pays 5% of her net income per month into a personal pension plan.
   b How much does Mrs Warren pay into the plan per month?

8 Table 2.11 gives information about people under
   35 years of age living in the UK in 1997 and 1998.
   Brackets give the total number of people
   younger than 35.

	Under 16	16–24	25–34	Total (millions)
1997	41%		32%	(28.4)
1998	42%		33%	(27.8)

   Table 2.11

   a Copy and complete the table.
   b Mark the following as 'true', 'false' or 'need more information'.
   From 1997 to 1998:   **i**   the proportion of under 35 year olds in the *16–24* year group decreased.
                        ii the number of under 16's increased
                        iii the total population of the UK fell.

9 An investment of £5000 grows by 6% during the first year, and by 7% during the second year.
   How much is the investment worth after two years?

LEVEL 3

10 Archive records of births, deaths and marriages
for a village are given in Table 2.12.
  a Calculate the population figure at the end of 1883.
  b Estimate the number of deaths in 1883.
    Give one reason why your answer may only be
    an estimate.
  c What percentage of the population married in 1883?
    (Take the population to be 5200.)

Population at end of 1882 = 5200 people
Population increase during 1883 = 4.5%
No. births in 1883 = 755
No. deaths in 1883 = _____
No. marriage ceremonies in 1883 = 195

Table 2.12

LEVEL 3 ## A multiplier can be used for repeated changes

Example

The population of a town is 25 800 (nearest hundred) and is expected to
increase by 4% each year. Estimate the population after 3 years.

Here's how...

The multiplier for each year is 1.04.
After 1 year expected population $= 1.04 \times 25\,800$
After 2 years expected population $= 1.04^2 \times 25\,800$
After 3 years expected population $= 1.04^3 \times 25\,800 = 29\,021$

Answer: 29 000 (nearest hundred)

hint
Population each year =
104% of the previous
year's population
104% = 1.04

hint
Give answer to the same
level of accuracy as
information given.

Example

A machine depreciates in value by 20% each year.
After how many years is its value reduced to about half of its original value?

Here's how...

The multiplier is 80% = 0.8
Using a calculator to repeatedly multiply by 0.8:
$0.8^2 = 0.64$ $0.8^3 = 0.512$ $0.8^4 = 0.4096$
The value closest to 0.5 has a power of 3. This gives the number of years.

Answer: Approximately 3 years

hint
100% − 20% = 80%

hint
Half is 0.5

Now try these...

1 Sam puts £200 into a building society account that gives 6% interest per annum. No other deposits or
   withdrawals are made. How much will there be in the account after 5 years?

2 A new car costs £12 000. It loses 15% of its value each year. How much is it worth after 4 years?

3 An investment plan offers 8% growth each year. £5799 is invested in the plan.
   a Assuming all interest is re-invested how much is the investment worth after:
     i one year   ii 4 years?
   b After how many years is the investment worth approximately double its initial value?

4 The height reached by a ball in each bounce is three-quarters of the height of the previous bounce.
   After how many bounces is the height reached by the ball approximately 10% of its normal height?

5 Wood expands when it gets wet. The diagram shows a 1.97 m by 0.73 m wooden front door. The metal door frame measures 2 m by 0.75 m. After a heavy storm, the sides of the door increase in length by 0.5% per hour.

a What are the dimensions of the door after 1 hour? (Give 5 d.p.)

b After Approximately how many hours will it become impossible to open the door?

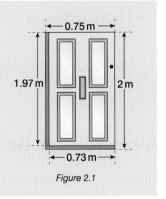

Figure 2.1

VAT is $17\frac{1}{2}\%$

Value Added Tax (VAT) is a tax on most goods and services. It is currently charged at $17\frac{1}{2}\%$. Prices without VAT are called pre-VAT prices. A pre-VAT price is increased by $17\frac{1}{2}\%$ to give the selling price.

Example

A microwave has a pre-VAT price of £240. Find the selling price.

Here's how...

VAT $= 17\frac{1}{2}\% \times$ £240

Using short cuts: 10% of £240 = £24

5% of £240 = £12

$2\frac{1}{2}\%$ of £240 = £6

VAT $=$ £24 + £12 + £6 = £42

Selling price $=$ £240 + £42

$=$ £282

Answer: Selling price = £282

hint

$17\frac{1}{2}\% = 10\% + 5\% + 2\frac{1}{2}\%$
5% is half of 10%.
$2\frac{1}{2}\%$ is half of 5%.

hint

Check:
Estimate of VAT
$= 20\% \times$ £240
$=$ £48

VAT	$= 17\frac{1}{2}\% \times$ **pre-VAT price**
$17\frac{1}{2}\%$	$= 10\% + 5\% + 2\frac{1}{2}\%$
selling price	$=$ **pre-VAT price + VAT**

hint

Find VAT on a calculator by multiplying by 0.175 (or 1.175 to include original cost)

Now try these...

1 A video recorder costs £160 without VAT. Find: **a** the VAT paid   **b** the selling price.

2 An emergency plumber charges a £35 call out fee and £10.25 per hour. A job lasts 4 hours.
   a Calculate the charge made by the plumber before VAT.
   b Calculate the total cost of the bill.

3 A Chinese takeaway offers a 10% discount to customers who collect their own orders. Prices on the menu are shown without VAT. Michelle places this order and collects the order herself.
   a How much is Michelle charged before VAT is added?
   b Calculate the total cost of the food (work to the nearest pence).

Dish	Quantity	Price
Pancake roll	3	£1.10
Kung po prawns	1	£4.20
Sweet & sour pork	1	£2.90
Egg fried rice	3	£1.50
Rice noodles	2	£2.55

Table 2.13

4 In this conference bill VAT is not charged on room hire.
Calculate:
a the VAT paid
b the total cost of the bill.

Room hire	£150 (no VAT)
Lunch	12 × £6.50 + VAT
Afternoon tea	10 × £0.60 + VAT

Table 2.14

A change can be expressed as a percentage

Percentage change measures how much a quantity changes compared to its **original value**. The change is either an **increase** or **decrease**.

Example

Sales in 'Ma's Bar' café in January were £3200. In February, sales fell to £2560. Find the percentage decrease in sales.

Here's how...

Decrease in sales $= £3200 − £2560$

$= £640$

% decrease $= \dfrac{\text{decrease}}{\text{original}} \times 100\%$

$= \dfrac{640}{3200} \times 100\%$

$= \dfrac{1}{5} \times 100\%$

$= 20\%$

Answer: Sales have fallen by 20%

hint

Change = larger amount − smaller amount

hint

% change is measured in terms of the **original** amount.

checkpoint

$\dfrac{640}{3200} = \dfrac{64}{320} = \dfrac{1}{5}$

⚠️ % change $= \dfrac{\text{change}}{\text{original}} \times 100\%$

Now try these...

1 Calculate the % change in the following amounts:
a original = £25, increase = £5   **b** original = £40, decrease = £16   **c** original = £40, final = £55

2 A £50 investment increases to £65. Calculate the % increase in the investment.

3 Because he is the only occupant, John's council tax bill is reduced from £70 to £52.50 per month. What is the % decrease in the bill?

4 The number of people celebrating New Years Eve in 1998 in a city was estimated to be 8 million. One year later, the figure was about 10 million.
a Find the % increase from 1998 to 1999.
b Give one reason for the large % increase of people in 1999.

5 Family ticket sales for a tourist attraction fell from 2596 in August to 1281 in September.
a Find the approximate % decrease in sales.
b Give one reason why Family ticket sales fell in September.

6 Table 2.15 shows sales figures for the
Whistle & Flute clothing store.
Mark the following as true or false:

Month	April	May	June
Sales (in £1000)	10	16	22

Table 2.15

 a There has been a steady increase in sales each month.
 b The % increase in sales has remained the same each month.
 c Sales have more than doubled from April to June.

7 The number of deaths on building sites in 1998 was 90. This represents a 20% increase on 1997.
What was the number of deaths on building sites in 1997? Write down the correct answer from:
 A 72   **B** 75   **C** 108

8 Denise is an athlete. Her performance is measured by the time it takes her to run 5 km.

> **Training Report**
>
> **Day 1:** Distance = 5 km   **Day 2:** Distance = 5 km
>   Time = 15 minutes   Time = 16 minutes
> Summary: Denise's times indicate an approximate 7% improvement in performance.

 a Find the % increase in times from Day 1 to Day 2.   **b** What is wrong with the summary?

LEVEL 3

9 Table 2.16 shows rainfall (in mm) over consecutive
years in three countries:

Which country experienced the highest:
 a rainfall each year?
 b % increase in rainfall between 1991 and 1992?
 c % change in rainfall between 1990 and 1992?

Country	Year 1990	Year 1991	Year 1992
England	998	784	866
Scotland	1523	1460	1602
Northern Ireland	1209	1009	1214

Table 2.16

2.1.4 Working with large and small numbers

One thousand is 1000 and one million is 1 000 000. The word 'billion' has had different meanings in the UK
and the USA, but it is now generally taken to mean 1 000 000 000 (i.e. a thousand million).

> **Example**
> Write in figures   **a** quarter of a million   **b** £6.75 billion
>   **c** ten million, twenty thousand, nine hundred and four
>
> **Here's how...**
> **a** $\frac{1}{4} \times 1\,000\,000 = 250\,000$ **Answer:** 250 000
> **b** $6.75 \times 1\,000\,000\,000 = 6\,750\,000\,000$ **Answer:** £6 750 000 000
> **c** Put in extra zeros where necessary.   **Answer:** 10 020 904

hint
> $\frac{1}{4}$ = 0.25. See ToTT.3 for decimal calculations.

LEVEL 3   Very large and small numbers are easier to work with in standard form

For example, the mass of the earth is 5 980 000 000 000 000 000 000 000 kg.
In standard form this is 5.98×10^{24} kg.

 **Numbers in standard form consist of a number between 1 and
10 multiplied by a power of 10. Positive powers of 10 mean
the number is large. Negative powers of 10 mean it is small.**

To write the number as an ordinary number, move the decimal point the number of places given by the power.

Here are some standard form numbers and their meanings:

Standard form	2.8×10^3	2.8×10^{-3}	1.25×10^5	1.25×10^{-5}	7×10^9
Ordinary number	2800	0.0028	125 000	0.000 012 5	7 000 000 000

Table 2.17

To enter standard form numbers into your calculator use the [EXP] key.

For example, to enter 7×10^9 press [7] [EXP] [9] **NB DO NOT ENTER THE × 10**

Example

Divide 4.2×10^7 by 3.5×10^{-4}

Here's how...

Press [4] [.] [2] [EXP] [7] [÷] [3] [.] [5] [EXP] [4] [+/−] [=]

The calculator display may give 1.2E11 or 1.2^{11}

Always write the answer correctly in standard form by writing in the 10:

Answer: 1.2×10^{11}

hint

Some calculators have a [(−)] key instead of [+/−]

Now try these...

1 Write in figures:
  a half a million   b £4.6 million   c nineteen thousand
  d eight billion   e £3.25 billion   f two and a quarter billion
  g five million, three hundred and fifty thousand and twelve
  h four hundred and fifty million, sixty-eight thousand and five

2 These numbers are in standard form. Write them as ordinary numbers.
  a 1.6×10^6 b 1.6×10^{-6} c 4×10^8 d 4×10^{-8} e 7.14×10^5 f 7.14×10^{-5}

3 Write these numbers in standard form.
  a 62 000 000   b 91 300   c 800 000 000   d 0.000 85   e 0.000 005   f 0.001 586

4 The areas of three of the continents are given below in square miles to 2 significant figures.
    Asia 1.7×10^7 North America 9.4×10^6 South America 6.9×10^6
  Which of these continents is: a the smallest   b the largest?

5 The masses of a proton, neutron and electron are given below. List them in increasing order of mass.
    Proton 1.672×10^{-24} g   Neutron 1.675×10^{-24} g   Electron 9.108×10^{-28} g

6 Use your calculator to work these out. Give your answers as ordinary numbers.
  a $7.6 \times 10^3 + 9.3 \times 10^2$ b $7 \times 10^{-2} - 3.2 \times 10^{-3}$ c $3.5 \times 10^4 \times 2 \times 10^{-3}$
  d $(6 \times 10^{-2}) \div (1.2 \times 10^3)$ e $(3 \times 10^2)^3$ f $\sqrt{6.4 \times 10^{-3}}$

7 Use your calculator to work these out. Give your answers in standard form.
  a $8 \times 10^{-7} + 1.2 \times 10^{-6}$ b $2.6 \times 10^9 - 7.3 \times 10^8$ c $4.75 \times 10^{-4} \times 2.56 \times 10^{-8}$
  d $(8.1 \times 10^6) \div (2.7 \times 10^{-3})$ e $(2.5 \times 10^{-3})^2$ f $\sqrt{4.9 \times 10^{13}}$

8 a The distance between the sun and the earth is 150 000 000 000 metres. Write this in standard form.
  b Light travels at a speed of 299 800 000 metres per second. Write this in standard form.
  c Use the formula: $\text{Time} = \dfrac{\text{Distance}}{\text{Speed}}$ to find the time it takes for light from the sun to reach the earth.
     Give your answer to the nearest second.

9 A telephone company with 2.4 million customers estimates that it deals with 1.5 billion calls per year.
   a Write this information in standard form.
   b Find the average number of calls per customer per day.

10 The table gives information about the workforce in 1997.

   a Calculate the number of employed 25–44 year-old **i** males **ii** females.
   b What percentage of the employed work force is aged 25–44?
   c What simple check can be made to see if your answer is correct?

Year: 1997	Employment total	% employed aged 25–44
Males	1.60×10^7	8.1%
Females	1.27×10^7	6.4%

Table 2.18

2.1.5 Working with negative numbers

Negative numbers are used for quantities that are **less than** zero, e.g. $-5°C$ means 5 degrees **below** zero. Problems involving negative numbers can often be solved using a number line.

Example

Adam makes a deposit into his bank account on Tuesday and withdraws some money on Wednesday. The balance of his account at the end of each day was:
Mon $-£200$ Tues £140   Wed $-£50$
How much did Adam: **a** deposit on Tuesday,
                     b withdraw on Wednesday?

Here's how...

This number line shows the balance at the *end of each day*.

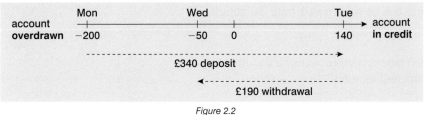

Figure 2.2

Answer: a £340   **b** £190

hint
A negative balance means it is *overdrawn*. (Adam owes the bank money.)

hint
For more negative number methods see ToTT.6.

checkpoint
Check:
£340 − £190 = £150.
There should be £150 more in the account at the end of Wednesday. From −£200 to −£50 is an increase of £150.

Now try these...

1 The temperature in Prague at midnight is 4.5°C. During the night, the temperature falls by 7°C.
   a What is the temperature in Prague by morning? (Sketch a thermometer to help if you wish.)
   b By midday, the temperature has risen by 2°C. What is the temperature in Prague at midday?

2 The sketch shows an iceberg in the sea. What is the height of the iceberg from base to tip?

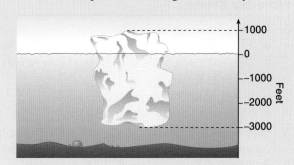

Figure 2.3

3 The temperatures (in °C) of three chemicals
   A, B and C are changing over time.
   a Copy and complete the table.
   b Which chemical has the lowest
     final temperature?

	Starting temperature	Temperature change	Final temperature
A	5	−12	
B	−4		6
C		−5	−12

Table 2.19

4 The bank balances of three customers A, B and C are: A −£382   B −£90.50   C £205.75
   a Who is the most overdrawn customer?
   b Customer A deposits £500 into his account. What is customer A's new balance?
   c Customer B withdraws £150 from his account. What is customer B's new balance?
   d A fourth customer, D, has £350 less than customer C. What is customer D's balance?

2.1.6 Working with units

There are many different units of time

For example: year, month, week, day, hour, minute, second.
The table shows the connection between these units:

1 year = 12 months	1 month ≈ 4 weeks	1 day = 24 hours
= 52 weeks		1 hour = 60 minutes
= 365 days		1 minute = 60 seconds

Table 2.20

hint

A more accurate result is
1 year = $365\frac{1}{4}$ days.
Most calendar months
are longer than 4 weeks.
1 year ≈ 12 × 4
= 48 weeks is *not* an
acceptable approximation.

A time period such as 90 minutes can be written as 1 hour 30 minutes.

> **Example**
>
> 100 minutes of a 3 hour video tape is used to record a documentary.
> How much tape is left in hours and minutes?
>
> **Here's how...**
>
> Convert 3 hours to minutes: Connection   1 hour = 60 minutes
> 3 hours = 3 × 60 = 180 minutes
> Time left on tape = 180 mins–100 mins
>   = 80 mins
> 80 mins = 60 mins + 20 mins = 1 hour and 20 mins
>
> **Answer:** 1 hour 20 mins

hint

Common sense says there
will be more minutes than
hours. *Multiply* by 60.

Now try these...

1 Convert the following:
   a 5 hours into minutes   b 3 years into months   c 3 years into weeks   d 132 months into years
   e 264 hours into days   f 91 days into weeks   g 1 hour into seconds   h 2 days into minutes

2 Convert the following:
   a 110 minutes into hours and minutes   b 42 months into years and months
   c 6480 minutes into days and hours

3 Jane wants to know how many weeks there have been in the last 2000 years. Which of the following
   calculations will give Jane the best approximation?
   A 2000 × 52   B 2000 × 12 × 4   C 2000 × 365 ÷ 7

4 The human heart beats roughly 100 000 times a day. Is this more than or less than one beat per second?

5 Rachael tapes these programmes:

The machine starts with a 4 hour blank tape.
Will there be enough room left to record two 40 minute documentaries?

Title	Duration
Prime Suspect	75 minutes
The Brit Awards	1 hour 35 mins

Table 2.21

Times can be given using the 12-hour or 24-hour clock

The 12-hour clock divides a day into two halves; morning (am) is from midnight to midday and afternoon (pm) is from midday to midnight.

The 24-hour clock runs continuously from midnight to midnight.

Example

Express as 12-hour and 24-hour clock times:
a half past nine in the morning   **b** quarter past five in the afternoon

Here's how...

a half past nine in the morning   = **9:30 am** (12-hour clock)
                                        = **0930** (24-hour clock)
b quarter past five in the afternoon = **5:15 pm** (12-hour clock)
                                        = **1715** (24-hour clock)

hint

5 + 12 = 17 hours
after midnight

Now try these...

1 Express the following as 12 and 24-hour clock times:
   a Quarter past eleven in the morning   **b** Half past two in the afternoon
   c Ten past three in the morning   **d** Quarter to seven in the morning
   e Twenty to eight in the evening   **f** Five to midnight

2 Write these as 24-hour clock times: **a** 7:18 am   **b** 4:21 pm   **c** 9:54 pm   **d** 12 noon

3 Write these as 12-hour clock times: **a** 1120   **b** 1353   **c** 0347   **d** 2222

To calculate journey times use hours and minutes

hint

Take care when time is
given as a decimal
number.

Example

A postman starts his round at 6 am. It takes him 3.1 hours to finish.
What time does he complete his round?

Here's how...

3.1 hours = 3 hours + 0.1 hours   and 0.1 hours = 0.1 × 60 mins = 6 mins
3 hours after 6 am is 9 am   6 minutes past 9 am is 9:06

Answer: He finishes at 9:06 am

To work out a journey time, break the journey into sections, working forwards in time to the nearest hour.

Example

A train leaves Paddington at 0920. It arrives in Swansea at 1250 the same day. How long did the journey take in hours and minutes?

Here's how...

section	period	
	hours	mins
0920 → 1000		40
1000 → 1200	2	
1200 → 1250		50
journey time =	2 hours	90 mins
=	2 hours +	1 hour + 30 mins
=	3 hours 30 mins	

Answer: 3 hours 30 mins

hint
Work forwards in time.

hint
90 minutes is longer than 1 hour.

Now try these...

1   A milkman starts his round at 5:15 am. It takes 4.6 hours to complete the round.
    At what time does he finish the round?

2   Michael leaves Leicester at 8:50 am for Manchester. The journey takes 2.2 hours.
    At what time does Michael arrive in Manchester?

3   An Apollo spacecraft is launched on Sunday at noon. It takes 4.3 days for the craft to arrive on the moon.
    a On which day does the craft reach the moon?
    b At what time does the craft touch down? Give your answer in 24-hour clock time.

4   A driver sets out from London at 3:20 pm. She drives for $2\frac{1}{2}$ hours, stops for a 25 minute break and then drives for another $1\frac{3}{4}$ hours to her destination. At what time does she reach her destination?

5   A private jet bound for Greece is due to leave at 0730. The jet takes off after a 40 minute delay and the flight lasts 3 hours 5 minutes. Greek time is two hours ahead of UK time. (When it is 1200 in the UK, it is 1400 in Greece.) At what local time does the jet land in Greece?

6   A coach left Victoria Station at 1935 and arrived in Dover at 2212 the same day.
    How long did the journey take?

7   An express train left Paddington at 1035 and was due to arrive in York at 1337.
    During the journey, the train was delayed and arrived in York at 1410.
    a How long was the delay?   **b** How long did the journey take?

8   A ferry is due to leave Dover at 2343. Passengers board the ferry half an hour before its departure time. The ferry arrives in Calais at 0117 the next day. Passport checks require the passengers to wait 35 minutes before they can leave the ferry. For how long in total are the passengers on the ferry?

9   A lorry driver sets out from his depot in London at 1128. He arrives in Kent at 1352.
    a How long does the journey take?

    After delivering his load, the driver leaves Kent at 1635 and heads back to London. He estimates that he must add 15% to his journey time to allow for London rush hour traffic.
    b At what time does he expect to return to his depot? Give your answer in 24 hour clock time.

LEVEL 3

10 Gary works in a factory. His basic rate of pay is £6.20 per hour.
    Table 2.22 shows Gary's time sheet for a week:
    Gary earns $1\frac{1}{2}$ times the basic rate for working after 6 pm.
    a How much does Gary earn per hour after 6 pm?
    b How much does Gary earn in this week?

Day	Clocks on	Clocks off
Monday	0900	1730
Tuesday	0845	1410
Wednesday	0720	1645
Thursday	1100	2030
Friday	0800	1615

Table 2.22

Timetables usually use the 24-hour clock

Each column refers to a particular journey. Read *down* a column for information on arrival times at each station or stop.

Example

The table shows part of a local train service timetable.

Station								
Rostock	0830	0845	—	0910	0920	0925	—	0955
Westway	0840	0855	0900	—	0930	—	0950	1005
Borchester	0855	0910	—	—	0945	—	1005	1020
Rochester	0901	0916	0925	0935	0951	—	1011	1026
Ant Caves	0913	0928	0937	—	1003	0958	1023	—
Carlow*	0922	0937	0946	—	1012	1007	1032	—
Hustings	0940	—	1004	0951	1030	1025	1050	1045
St Par's	0952	—	1016	1003	—	1037	1102	1057
Moss Bay	1007	—	1031	1018	—	1052	1117	1112

*mainline connection

Table 2.23

hint

— means the train does not stop at the station.

Mr Robbins catches the 0840 from Westway.
Calculate his journey time to Moss Bay.

Here's how...

Arrival time at Moss Bay = 1007

section	period	
	hours	mins
0840 → 0900		20
0900 → 1000	1	
1000 → 1007		7
journey time	= 1 hour +	27 mins

Answer: Journey time = 1 hour 27 minutes

Now try these...

Questions 1 and 2 refer to the timetable in Table 2.23 above.

1 **a** Mary catches the 0950 at Westway. Calculate her journey time to Carlow.
   b Jacob boards the 0925 at Rostock. Calculate his journey time to St Par's.
   c Judith catches the 0937 at Ant Caves. A signal failure delays her arrival into Moss Bay by 25 minutes. Calculate her journey time to Moss Bay.

2 **a** Yolanda arrives at Rostock Station at 0850. She catches the next train to Hustings.
      At what time does she arrive at Hustings?
   b Spike arrives at Westway Station at 0845. He catches the next train to St Par's.
      At what time does he pass through Carlow?
   c Trisha lives in Rostock. She needs to catch the 1017 mainline train from Carlow.
      What is the latest train she can catch at Rostock?
   d Ben buys a one day travel card at Borchester Station. The card is not valid until after 9:30 am.
      What is the earliest time he can arrive in Moss Bay?

Metric units are decimal measures

Metric units include:

Length: kilometre (km), metre (m), centimetre (cm),
        millimetre (mm)
Mass: tonne (t), kilogram (kg), gram (g), milligram (mg)
Capacity: litre (*l*), centilitre (c*l*), millilitre (m*l*)
(Capacity of solid objects is covered in Section 2.3)

Length	Mass	Capacity
1 km = 1000 m	1 t = 1000 kg	1 *l* = 100 cl
1 m = 100 cm	1 kg = 1000 g	1 cl = 10 ml
1 cm = 10 mm	1 g = 1000 mg	(1 *l* = 1000 ml)

Table 2.24 shows how to convert between the units

Example

Convert   **a**   3 metres into centimetres   **b**   7500 grams into kilograms

Here's how...

a   Connection: 1 m = 100 cm $3 \times 100 = 300$ cm

Answer: 300 cm

b   Connection: 1 kg = 1000 g $7500 \div 1000 = 7.5$ kg

Answer: 7.5 kg

checkpoint
Common sense says there will be more centimetres than metres.

checkpoint
Common sense says there will be fewer kilograms than grams.

Example

An empty tank with a capacity of 7.4 *l* is filled with water using a 80 c*l* jug.
Find the minimum number of times the jug is used.

Here's how...

Connection: $1\,l = 100\,cl$
Convert 7.4 *l* into centilitres $7.4 \times 100 = 740\,cl$
Divide the capacities $740 \div 80 = 74 \div 8 = 9.25$

Answer: 10 times

hint
Assume the jug is full each time.

hint
Although 9.25 is closer to 9 than 10, the jug is used 10 times in total.

It is useful to have a rough idea of the size of some everyday objects. This helps in estimating the size of other objects and judging whether answers from calculations are sensible. The tables give some examples.

Object	Length
Thickness of 1p coin	1 mm
Width of little finger	1 cm
Length of standard ruler	30 cm
Length of long stride	1 m
Length of small car	3 m
Length of football pitch	100 m

Table 2.25

Object	Capacity
Teaspoon	5 m*l*
Glass of wine	150 m*l*
Mug	30 c*l*
Standard bottle of wine	70 c*l*
Bucket	5 litres
Bath	300 litres

Table 2.26

Object	Mass
£1 coin	10 g
Bag of crisps	25 g
Bag of sugar	1 kg
Adult	70 kg
Car	1 tonne
Locomotive	100 tonnes

Table 2.27

Measurements are always estimates. The accuracy that is possible depends on the sensitivity of the instrument and the scale used. The length of this paperclip is between 2.5 and 2.6 cm. It is nearer to 2.6 cm and can be given as ***2.6 cm to 1 decimal place***.

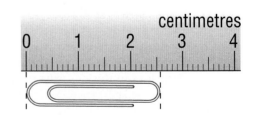

Figure 2.4

Now try these...

1 Convert the following:
   a 3 kg into grams   **b** 25 *cl* into millilitres   **c** 7500 m into kilometres   **d** 340 *cl* into litres

2 Convert the following
   a 5 km into centimetres   **b** 3.2 tonnes into grams   **c** 0.15 litres into millilitres

3 A shopping basket contains the following items:

     450 g piece of steak, 2.4 kg whole chicken, 3 × 400 g tins of tomatoes

   Find the total weight of the items in: **a** grams   **b** kilograms.

4 A lift holds a maximum of 0.75 tonnes. How many people weighing 60 kg are allowed in the lift?

5 A screw is used to fix a hinge to a wooden door. Each whole turn of a screwdriver advances the screw 5 mm into the wood. How many turns are needed to advance the screw 3.5 cm?

6 A sack of potatoes are placed on a pair of scales.

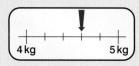

Figure 2.5

   How much does the sack weigh in:

   a kilograms   **b** grams?

7 A tablespoon has a capacity of 15 ml.
   The number of whole tablespoons of oil that can be taken from a $\frac{1}{4}$ litre bottle of olive oil is:
   A 16   **B** 17   **C** 167

8 The towns of Archdale and Bedwyn are 15 km apart. A new road is being built to join the two towns.
   45 metres of road is laid each day.
   a What percentage of the job is complete after 20 days?
   b At this rate how many days will it take for the job to be completed?

9 Estimate:
   a the thickness of a £1 coin,   **b** the width of a sheet of A4 paper   **c** the length of a tennis court.

10 Estimate:
   a the capacity of a washing-up bowl   **b** the mass of a cat
   c the mass of a CD (without its case).

11 Estimate the length of these objects. Give your answers to 1 decimal place.

a
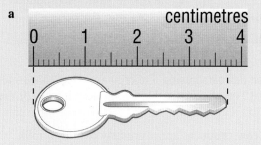

b
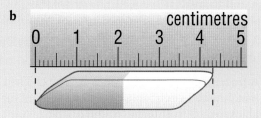

Figure 2.6

Imperial units were used in this country before metric units

Imperial weights and measures are still commonplace. For example, you probably know your height in feet and inches (imperial) rather than metres (metric).

Imperial units include:

Length: mile, yard (yd), foot (ft or '), inch (in or '')

Mass: ton, stone (st), pound (lb), ounce (oz)

Capacity (liquid): gallon, pint, fluid ounce (fl. oz)

Table 2.29 shows how to convert imperial units:

Length	Mass	Capacity
1 mile = 1760 yds	1 ton = 160 st	1 gallon = 8 pints
1 yd = 3 ft	1 st = 14 lbs	1 pint = 20 fl. oz
1 ft = 12 in	1 lb = 16 oz	

Table 2.28

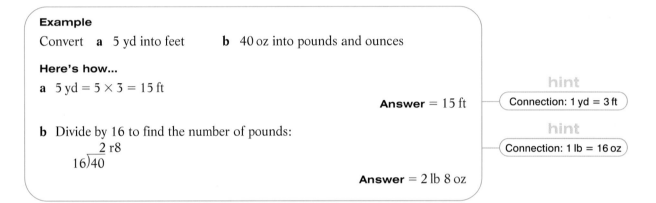

Example

Convert   **a**   5 yd into feet   **b**   40 oz into pounds and ounces

Here's how...

a   5 yd = 5 × 3 = 15 ft

Answer = 15 ft

hint
Connection: 1 yd = 3 ft

b   Divide by 16 to find the number of pounds:

$$16\overline{)40}\;\;\;\;2\;r8$$

Answer = 2 lb 8 oz

hint
Connection: 1 lb = 16 oz

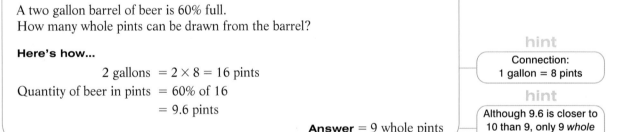

Example

A two gallon barrel of beer is 60% full.
How many whole pints can be drawn from the barrel?

Here's how...

2 gallons = 2 × 8 = 16 pints

Quantity of beer in pints = 60% of 16

= 9.6 pints

Answer = 9 whole pints

hint
Connection:
1 gallon = 8 pints

hint
Although 9.6 is closer to 10 than 9, only 9 *whole* pints can be drawn from the barrel.

Now try these...

1   Convert:
   a 4 ft into inches   **b** 3 st into pounds   **c** 26 lb into stones and pounds
   d $4\frac{1}{2}$ pints into fluid ounces   **e** 36 pints into gallons and pints **f** $3\frac{1}{2}$ tons into stones

2   A bag of apples weighs 3 lb 9 oz. A bag of pears weighs 4 lb 8 oz. How much do the two bags weigh altogether in pounds and ounces?

3   A frozen chicken weighs 3 lb 8 oz. The chicken is to be defrosted using a microwave. Guidelines state that 5 minutes defrosting should be allowed for every 10 oz of chicken. How long does it take for the chicken to defrost using the microwave?

4   A farmer's milk churn can hold 13 gallons of milk. A cow produces approximately 6 pints of milk each milking session. During one session, the empty churn is completely filled with milk. Estimate the number of cows milked.

You can convert between metric and imperial units

Job application forms for manual work often ask for heights to be given in metres. Applicants who only know their height in feet and inches will need to convert from imperial to metric units.

The table shows some *approximate* imperial and metric conversions.

Length	Mass	Capacity
5 miles ≈ 8 km 1 ft ≈ 30 cm 1 in ≈ 2.5 cm	1 kg ≈ 2.2 lbs	$1\,l \approx 1\frac{3}{4}$ pints

Table 2.29

Example

A supermarket shows imperial measurements alongside their metric equivalents. Find the missing quantities.

    Potatoes   5 kg   ≈ _____ lb

    Orange juice _____ litres ≈ $3\frac{1}{2}$ pints

Here's how...

Potatoes: $5 \times 2.2 = 11$

Answer: 5 kg ≈ 11 lb

> *hint*
> Connection: 1 kg ≈ 2.2 lbs

Orange juice: $3\frac{1}{2} \div 1\frac{3}{4} = \frac{7}{2} \div \frac{7}{4}$

$$= \frac{7}{2} \times \frac{4}{7}$$

$$= 2\,l$$

Answer: $3\frac{1}{2}$ pints $\approx 2\,l$

> *hint*
> Connection:
> 1 litre ≈ $1\frac{3}{4}$ pints

Now try these...

1 Convert the following quantities:
  a 3 ft into centimetres   **b** 11 kg into pounds   **c** 6 in into centimetres   **d** $12\,l$ into pints

2 Convert the following quantities:
  a 60 cm into feet   **b** 25 cm into inches   **c** 32 km into miles   **d** 1 st 8 lb into kilograms

3 Which of the following quantities is the *greater*?
  a 30 miles or 50 km   **b** 9 ft or 250 cm   **c** $20\,l$ or 38 pints   **d** 15 in or 37 cm

4 Which weighs more: 1 ton of feathers or 1 tonne of coal? (hint: convert 1 ton into lbs)

5 A petrol tank holds $10\frac{1}{2}$ gallons. Find its capacity in litres.
  (Remember 1 gallon = 8 pints.)

6 Wooden picture frames are 20 cm by 25 cm. Glass is bought to place in the frame.
  Which size glass should be bought?
  Write down the correct answer from: **A** 8″ by 11″   **B** 9″ by 10″   **C** 8″ by 10″

7 1 ft is slightly more than 30 cm. A telephone cable is to run around the edge of a room from point A to point B. Which is the least length of cable that should be bought? Choose the correct answer from:
  A 5.4 m   **B** 5.8 m   **C** 6 m   **D** 6.1 m

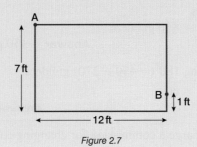

Figure 2.7

LEVEL 3

8 A bathroom wall is being decorated with 4″ square tiles.
   The dimensions of the wall are 0.95 m by 0.8 m.
   a How many whole tiles can be placed along the first row?
   b Calculate the area of a tile in cm².

   Tiles can be cut in half. Each half can be used to complete any gaps.
   c How many tiles must be bought to completely tile the wall?

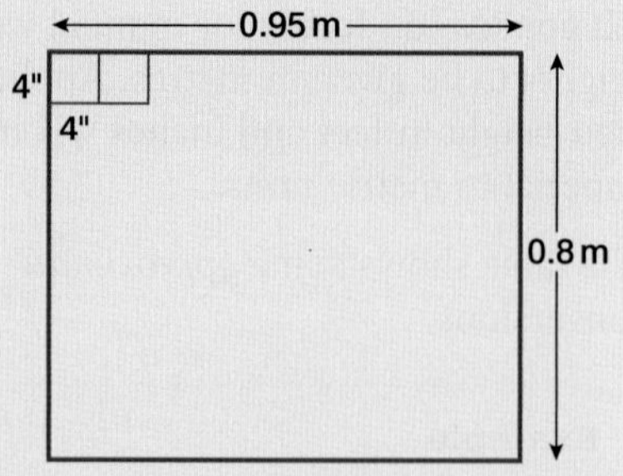

Figure 2.8

9 A prisoner has escaped from the UK and is heading to the USA.
   British police send recent details of the criminal to the FBI:
   Height = 1.73 m   Weight = 75 kg
   The FBI arrest four suspects.
   Which of the suspects most closely resembles the escaped prisoner in height and weight?

Suspect	Height (ft and in)	Weight (st and lb)
Joe 'Mad' Maddoc	5′7″	10 st 5 lb
Jack 'Knife McClain	5′9″	11 st 8 lb
PG 'Tea Leaf' Tipps	5′11″	11 st 3 lb
'Poisonous' Pete Potter	5′9″	11 st 11 lb

Table 2.30

Each country has its own currency

An *exchange rate* describes
the value of one currency
in terms of another.
Table 2.31 shows the
exchange rates for
various countries:

Country	Currency	Exchange rate (£1 =)
Austria	Schillings	21.4
Belgium	Francs	63
France	Francs	10.2
Germany	Marks	3.1
Italy	Lire	3035
Ireland	Punts	1.2
Netherlands	Guilders	3.0
Spain	Pesetas	260
Sweden	Kronor	13.2
Switzerland	Francs	2.4

Table 2.31

hint

Exchange rates change
daily. Most newspapers
and travel agents advertise
the daily rates.

Example

Brian changes £150 into guilders. He spends 270 guilders in the Netherlands.
He changes the rest into £s when he gets home. How much:
a in guilders does Brian take to the Netherlands?
b in £s is left after the holiday?

Here's how...

From Table 2.32 £1 buys 3 guilders
a Amount in guilders   3 × £150 = 450 guilders

                                       Answer = 450 guilders

b Brian spends 270 guilders, leaving 180 (= 450 − 270) guilders.

   Amount in £s   180 ÷ 3 = 60

                                       Answer = £60

hint

Brian is going abroad,
so multiply by the
exchange rate.

hint

Brian is returning home,
so divide by the
exchange rate.

In practice, banks and travel agents charge a commission for changing currency.
The commission is often removed before the currency is changed.

Now try these...

Use Table 2.31

1 Amanda is returning to the UK with 756 Belgian francs.
   How much will Amanda have if she exchanges it for pounds?

2 Jane changed £100 into French francs. She spends 820 francs.
   a How many French francs did she receive?   b How many French francs does she have left?

3 Mike changes £200 into pesetas for his holiday in Spain.
   a How many pesetas does he get?
   b If Mike spends 39 000 pesetas, how much in £ will he have left?

4 Tom wants to change £150 into guilders. The bank charges Tom 2% commission.
   a How much can Tom change into guilders?   b How many guilders does Tom receive?

5 7400 francs are transferred to a UK bank account. The amount received is less than £500. From
   which country was the money transferred?
   Write down the correct answer from: **A** France   **B** Switzerland   **C** Belgium

6 Omar changes £850 into German marks. He spends 2175 marks.
   The remaining marks are changed into £ with which Omar opens a bank account.
   Which of these calculations gives the amount in Omar's account?
   A $850 \times 3.1 - (2175 \div 3.1)$ **B** $850 - (2175 \div 3.1)$ **C** $(850 \times 3.1) - (2175 \times 3.1)$

LEVEL 3

7 Chris is back-packing across Europe. He changes £540 into guilders.
   a How many guilders does Chris get?

   After spending 620 guilders, Chris travels to Belgium. He changes all his guilders into francs.
   b How many Belgian francs does Chris receive?

8 Guiseppe wins the Italian National Lottery jackpot of 380 million lira. Guiseppe transfers the money
   to a UK bank account. Would Guiseppe be classed as a millionaire in the UK?

9 Toni has planned a five week holiday. She is spending the first three weeks in Switzerland and the last
   two weeks in Austria. Her spending money for the trip is £1850. A travel agent charges the following
   rates of commission per transaction:

   > Up to £1000 exchanged – commission = £5
   > Over £1000 exchanged – commission = 0.5%

   The agent only supplies whole units of any currency.

   a How much does Toni pay in commission?

   Toni decides to split her spending money between Swiss francs and Austrian schillings in the ratio of
   the time she will spend in each country.

   b How many: **i** Swiss francs   **ii** Austrian schillings   does Toni take with her?

   On holiday Toni spends $\frac{3}{4}$ of each type of currency. When she returns, she is charged 1% commission
   for exchanging her foreign currency back into pounds.

   c How much: **i** does Toni receive in £   **ii** has Toni spent in total?

Using Formulae

2.2.1 Formulae and substitution

A formula shows how two or more quantities are related

It can be given in words or using letters.

For example, an amount of money can be changed from pounds to Dutch guilders using the exchange rate £1 = 3 guilders. This means that £2 = 6 guilders, £5 = 15 guilders and so on. The number of guilders is three times the number of pounds. This can be written as a formula in words or using letters:

In words:	Using letters
number of guilders = 3 × number of pounds	$g = 3 \times p$ or $g = 3p$

> **hint**
> g is the number of guilders
> p is the number of pounds.
> The multiplication sign '×' can be omitted.

Suppose a plumber charges customers a fixed charge of £30 for coming to their house (the 'call-out' charge) plus £20 for each hour he spends on the work. This can be written as a formula:

In words:	Using letters
Charge in £ = number of hours × 20 + 30	$C = 20h + 30$

> **hint**
> C is the total charge,
> h is the number of hours.

The amount charged for a particular job lasting 4 hours can be found by replacing 'number of hours' or h by 4. This is called **substituting** h with the value 4.

In words:	Using letters
Charge in £ = 4 × 20 + 30	$C = 20h + 30$
= 80 + 30	= 20 × **4** + 30
Charge = **£110**	= 80 + 30
	Charge = £110

> **hint**
> Substitute $h = 4$

Sometimes formulae are more difficult, involving brackets, squares or other powers.

> **hint**
> This formula can be written in words:
> Largest mortgage
> = 2.5 × total gross annual income

Example

The formula $M = 2.5(I + J)$ gives the largest mortgage M (in £) available to a couple. I and J are the gross annual incomes (in £) of each person.
Ian and Jane want to buy a house.
Their gross annual incomes are: Ian £19 250   and   Jane £21 750
Calculate the largest mortgage Ian and Jane can get.

Here's how...

Substitute $I = $ **19 250** and $J = $ **21 750** into the formula:
$$M = 2.5 (I + J)$$
$$= 2.5 (\mathbf{19\,250 + 21\,750})$$
$$= 2.5 \times 41\,000$$
$$= 102\,500$$

Answer: Largest mortgage = £102 500

> **hint**
> Use BODMAS to work out the answer (see ToTT.3 and ToTT.7).

> **checkpoint**
> Using approximations:
> 3(20 000 + 20 000)
> = 3 × 40 000
> = 120 000

Example

A town is experiencing a flu epidemic.

The formula: $n = \dfrac{w^2}{4}$ is used to estimate the number n of people (in 1000's)

with flu where w is the number of weeks since the outbreak of the virus.

a How many people have flu 6 weeks after the outbreak?

b If the population of the city is 32 500 (to the nearest 100), explain why the formula does not give a sensible result for 12 weeks after the outbreak.

Here's how...

a Substitute $w = 6$: $\qquad n = \dfrac{w^2}{4} = \dfrac{6^2}{4} = \dfrac{36}{4}$

$\qquad\qquad\qquad\qquad\qquad = 9$

Answer: $n = 9000$

b When $\qquad w = 12$: $\qquad n = \dfrac{w^2}{4} = \dfrac{12^2}{4} = \dfrac{144}{4}$

$\qquad\qquad\qquad\qquad\qquad = 36$

Answer: The formula suggests 36 000 people will have flu after 12 weeks. This is not possible because the population of the city is only 32 500.

hint

Written in words: number of people with flu (in thousands) =

$\dfrac{\text{(number of weeks since outbreak)}^2}{4}$

Letters give a much neater formula!

hint

6^2 means 6×6.

hint

n is measured in 1000s.

Now try these...

1 The formula that gives the cooking time for a chicken is: time in minutes = $40 \times$ weight in kilograms + 20. Find the cooking time for a chicken weighing: **a** 2 kg **b** 1.7 kg.

2 The length of fabric needed to make a pair of curtains is given by the formula:

$\qquad$ length of fabric in metres = $2 \times$ (height of window in metres + 0.8)

Find the length of fabric needed for a window of height: **a** 2 m **b** 1.7

3 The formula $A = 18h + 25$ gives the amount charged in £ by a plumber for working h hours.
 a How much does the plumber charge for working: **i** 3 hours **ii** 4 hours?
 b What is the plumber's hourly rate of pay?
 c How much is the plumber's call-out fee?

4 An electrician charges at a rate of £16 per hour. Her call-out fee is £25. Which of these formulae gives the amount charged in £ by the electrician for working h hours?
 Choose the correct answer from: **A** $A = 25h + 16$ **B** $A = 25 + 16h$ **C** $A = 16(h + 25)$

5 The phone bill for a month, in which t minutes of calls are made, is given by the formula:

$\qquad B = 15 + 0.4t$ where B is the total bill in £ for the month.

 a Find the bill for a month in which 50 minutes of calls are made.
 b Line rental is the monthly charge paid even when no calls are made. How much is the line rental?

6 Sandra sells double glazing. Her earnings each day are found using the formula: $E = \dfrac{S}{10} + 6h$

Day	Sales made	No. hours worked
Monday	£400	6
Tuesday	£600	4
Wednesday	£550	5

Table 2.32

E = Amount earned in the day (£)
S = Total value of sales in the day (£)
h = No. of hours worked in the day
Table 2.32 shows her records for the beginning of one week.
 a On which day did Sandra earn the most?
 b On Thursday, Sandra earns £90 and works for 3 hours.
 The total sales figure for Thursday is: **A** £270 **B** £720 **C** £850

7   The formula F = 1.8C + 32 converts a temperature from Celsius (°C) into Fahrenheit (°F).
    a   On a particular day in July, the highest temperature in Rome is 30°C and in Naples is 82°F.
        Which city had the higher temperature?
    b   On a particular night in December, the lowest temperature in Rome is −5°C and in Naples is 25°F.
        Which city had the lower temperature?
    c   The highest temperature setting on gas ovens is about 550°F. A recipe says an oven is to be heated
        to 300 degrees. Is the recipe referring to 300°C or 300°F? Give a reason for your answer.

8   The exchange rate for converting UK pounds into German marks is £1 : 3DM. An exchange bureau
    charges £2 commission for converting amounts less than £500 into German marks. The commission
    is deducted from the amount before it is converted. Which formula gives the number of marks, m,
    received for exchanging £p (where p is less than 500)? Write down the correct formula from:
    A $m = 3p - 2$ **B** $m = 2p - 3$ **C** $m = 2(p - 3)$ **D** $m = 3(p - 2)$

9   The formula $C = \dfrac{5(F - 32)}{9}$ converts a temperature from Fahrenheit (°F) into Centigrade (°C).

    a   Convert: **i** 77°F   **ii** 86°F   into °C.

    Prina wants to convert 82°F into °C. She presses the following keys on her calculator:
    ⑤ ✕ ⦅ ⑧ ② ⊖ ③ ② ÷ ⑨ ⦆ ENTER   **392.22... °C**
    b   How can the answers to part **a** be used to show Prina has made a mistake?
    c   What mistake did Prina make? Write down the sequence of calculator keys she should have used.

10 The power in an electrical circuit is given by the formula $P = \dfrac{V^2}{R}$ where P is the power in watts, V is
    the voltage in volts and R is the resistance in ohms.
    Find the power when: **a** $V = 20$ and $R = 16$ **b** $V = 240$ and $R = 2000$.

LEVEL 3

11 If an amount £P is invested in a savings account that gives R% interest per annum, the amount in the
    account after n years is given by the formula $A = P(1 + 0.01R)^n$ where A is the amount in £.
    Calculate the amount in the account after 10 years if £2500 is invested at a rate of 6% per annum.

2.2.2 Compound measures

A **compound measure** relates two quantities.
For example, distance and time are related by the formula: Speed = Distance ÷ Time (see below).

To find speed divide distance by time

Suppose the speedometer of a car shows a constant reading of 50 kilometres per hour (kph)
At this speed the car will cover 50 km in 1 hour, 100 km in 2 hours, 150 km in 3 hours and so on.

The relationship between distance, speed and time can be written as a formula:
in words: ***Distance travelled = Speed × Time***
or using letters: $D = S \times T$

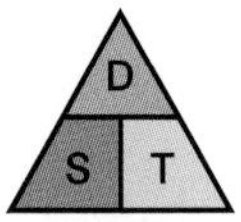

Putting the letters in a triangle can help you remember the relationship.
It can also be used to write the relationship in different ways.

Figure 2.9

To write down a formula for one of the letters, cover up that letter and what is
left gives you the rest of the formula.

To get a formula for time, cover up the T, giving $T = \dfrac{D}{S}$

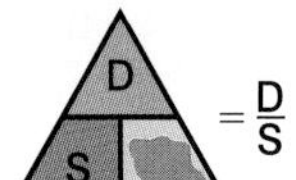

Figure 2.10

To get a formula for speed, cover up the S, giving $S = \dfrac{D}{T}$

The formula for speed can also be used to give the unit for speed.

If distance is in kilometres and time is in hours, then the speed is in $\dfrac{km}{h}$ i.e. km/h (or kph).

If distance is in metres and time is in seconds, then the speed is in $\dfrac{m}{s}$ i.e. m/s.

Example

A coach is travelling at a steady speed of 50 kph. Find:
a how far the coach travels in 12 minutes
b how long it takes the coach to cover a distance of 15 km.

Here's how...

a 12 minutes = 0.2 hours
To find a distance, cover up D in the speed triangle.

$$D = S \times T = 50 \times 0.2$$
$$= 10 \text{ km}$$

Answer $= 10$ km

b To find a time, cover up the T in the speed triangle.

$$T = \frac{D}{S} = \frac{15}{50}$$
$$= 0.3 \text{ hours}$$
$$= 0.3 \times 60 \text{ minutes}$$

Answer $= 18$ minutes

hint

Note the time is in minutes. Convert this to hours to be consistent with speed in kp**h** .

hint

To convert minutes to hours *divide* by 60:
12 minutes $= \frac{12}{60}$ hours
$= 0.2$ h

checkpoint

Using an inverse method:
10 km in 0.2 hours gives speed,
$S = \dfrac{D}{T} = \dfrac{10}{0.2} = \dfrac{100}{2} = 50$ kph
Correct!

hint

To convert hours to minutes *multiply* by 60.

NB When the speed is not constant average speed $= \dfrac{\text{distance}}{\text{time}}$.

Density is mass per unit volume

Figure 2.11

A 1 m³ block of steel weighs more than a 1 m³ block of wood. Steel is more dense than wood and so has a greater density. The triangle gives the relationship between mass, volume and density.

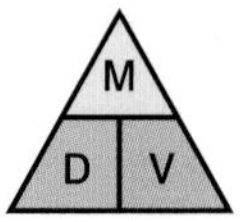

hint

Note density unit
$= \dfrac{\text{mass unit}}{\text{volume unit}} = \dfrac{kg}{m^3}$
$= kg/m^3$ (kg per m³)

Example

The density of oak timber is 720 kg/m³.
A section of an oak tree has mass 1080 kg.
Find the volume of the section.

Here's how...

To find a volume, cover up the V in the triangle:

$$V = \frac{M}{D} = \frac{1080}{720} = \frac{108}{72}$$

$$72\overline{)108.^{36}0} \quad \frac{1.5}{}$$

$$= 1.5 \text{ m}^3$$

Answer $= 1.5$ m³

checkpoint

Using an inverse method:
Mass $M = D \times V = 720 \times 1.5$
$= 1080$, correct!

Pressure is the force per unit area

The triangle gives the relationship between pressure, force and area. If the force is measured in pounds weight and the area is in square inches, the unit for pressure is pounds per square inch (abbreviated to psi).

Figure 2.12

Compound measures are often described as rates

One example is rate of pay.

Example

Mike's basic rate of pay is £7 per hour for a 35 hour week. Overtime is paid at 'time and a half'. How much does Mike receive for working 39 hours during one week?

Here's how...

Basic wage	35×7	$= £245$
Overtime rate	1.5×7	$= £10.50$ per hour
Number of hours overtime	$39 - 35$	$= 4$ hours
Payment for overtime	4×10.5	$= £42$
Total amount received	$£245 + £42$	$= £287$

Answer: Mike earns £287

hint

Formula for wage:
Amount earned = No. hours worked × Rate of pay

checkpoint

Using a *different* method: Each hour of overtime is worth $1\frac{1}{2}$ hours at the basic rate. Total number of hours at basic rate $= 35 + 4 \times 1\frac{1}{2} = 35 + 6 = 41$. Total wage $= 41 \times £7 = £287$

When chemicals are added to water the rate is often given in parts per million (ppm). For example, suppose chlorine is added to a swimming pool at a rate of 80 ppm. If the swimming pool holds 500 000 litres of water (i.e. half a million litres), the amount of chlorine needed is 40 litres (i.e. half of 80).

Now try these...

1 A car travels at a steady speed for 30 minutes. It covers a distance of 28 km.
   a At what speed is the car travelling? Give your answer in kilometres per hour.
   b How long, in minutes, will it take the car to travel 42 km?

2 A coach leaves London for Torquay. The distance from London to Torquay is 390 km. The coach travels at a steady speed of 80 kph for the first 240 km and at 60 kph for the rest of the journey. Find the total journey time to Torquay.

3 The density of concrete is 2300 kg/m³. The volume of a concrete block is 0.12 m³. Find its mass.

4 A force of 240 pounds weight acts on a surface whose area is 25 square inches. What is the pressure in psi?

5 Jim's map says the distance to Berlin is 155 miles. After travelling 10 minutes at a steady speed, his car passes a signpost that says the distance to Berlin is 228 km.
   a How far in kilometres has Jim travelled? (Use the conversion 5 miles = 8 km)
   b Give one reason why your answer to part **a** might not be accurate.
   c Estimate the speed of the car in kilometres per hour.
   d Assuming that the car travels at the speed found in part **c**, estimate how much longer the journey to Berlin will take. Write down the correct answer from:
      A 1.5 hours   **B** 1.9 hours   **C** 2.25 hours   **D** 3.8 hours

6 A lorry driver leaves Dover and drives at a steady speed to Norfolk. He has a one hour lunch break during which the lorry is unloaded. He then returns to Dover along the same route at a steady speed. Copy and complete the driver's time sheet:

Departure time	Distance	Speed	Journey time	Arrival time
London: 1000	240 km	120 km/h	....	Norfolk: ....
Norfolk:	....	....	....	London: 1612

Table 2.33

7 A solar panel made of sheet glass has mass 250 kg and volume 0.1 m³.
   a Find the density of sheet glass in kg/m³.
   Another solar panel made of the same type of glass has mass 600 kg.
   b Find the volume of this solar panel. (hint: Assume this panel has the same density as the first panel.)

8 An aeroplane preparing to land is descending at a steady rate of 50 feet per second.
   a How many feet does the aeroplane descend in half a minute?
   Before starting its descent, the aeroplane has an altitude of 12 000 feet.
   b How long in minutes does it take the aeroplane to land?

9 Chlorine is added to a paddling pool at a rate of 60 ppm. The paddling pool holds 10 000 litres of water. How much chlorine is needed?

10 Jan earns £6 per hour for a basic working week of 37 hours. Overtime is paid at 'time and a quarter'. How much does Jan earn in a week when she works 40 hours?

LEVEL 3 Convert components of compound units one at a time

Speed is often measured in kilometres per hour (kph) or miles per hour (mph).

Example
The speed limit on a road in France is 60 kph. Write this in mph.

Here's how...

8 km = 5 miles means 1 km = $\frac{5}{8}$ mile and 1 kph = $\frac{5}{8}$ mph

$60 \text{ kph} = 60 \times \frac{5}{8}$

$= \frac{300}{8} = \frac{75}{2}$

$= 37.5$

Answer = 37.5 mph

hint
Reduce to unity.

Example
A supersonic jet flies at 2400 kph. Convert this speed to metres per second.

Here's how...

In 1 hour, the jet flies 2400 km

Convert km to m: 2400 km = 2 400 000 metres
Convert metres per hour to metres per second in two stages:
Divide by 60:   2 400 000 ÷ 60 = 40 000 metres per *minute*
Divide by 60 again to find the speed in metres per *second*

$40 000 \div 60 = 666.\dot{6}$

$\approx 667 \text{ m/s}$

Answer = 667 m/s

hint
In 1 minute, the jet travels $\frac{1}{60}$ of the distance flown in 1 hour.

Now try these...

1   The strongest wind recorded in a hurricane blew at 176 kph. Convert this speed into miles per hour.

2   The *InterCity-125* train travels at 125 mph. Convert this speed into kilometres per hour.

3   A cannon ball is fired at a speed of 27 kph.
    a Find the speed of the cannon ball in metres per second.
    The barrel of the cannon is 1.5 metres long.
    b How long does it take for the ball to exit from the cannon?

4   In 1954, Roger Bannister ran a distance of 1 mile in just under 4 minutes.
    One newspaper reported that his average speed was about 24 kph.
    Is the report true? Show some calculations to support your answer.

5   The speed of light is approximately 3×10^8 metres per second.
    a Convert the speed of light to miles per second.
    The distance between the Sun and the Earth is approximately 9.3×10^7 miles.
    b If the Sun were to stop shining now, how many minutes later would we notice the effect on Earth?

6   A fighter jet flies at 1080 miles per hour.
    a Convert this speed to metres per second.
    b Estimate how long (in seconds) it took you to answer part **a**.
    c Estimate (in km) how far the jet has flown in the time it took you to find the answer to part **a**.

2.2.3 Working with equations

Suppose a shirt and tie cost £35. If the tie costs £7, how much does the shirt cost? The answer can be found by taking the cost of the tie from the total cost: £35 − £7 = £28.

An **equation** is a mathematical way of giving information. Letters are used to represent unknown values. Take the shirt and tie example above. If s is used to represent the cost of the shirt (in £), the total cost of the shirt and tie (in £) is $s + 7$. But the total cost is £35. This information can be used to write down an equation:

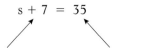

$$s + 7 = 35$$

Left-hand side (LHS)   Right-hand side (RHS)

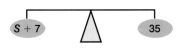

Figure 2.13

The LHS and RHS are equal. You can think of this as a balance.
Provided the same amount is added to or subtracted from both sides, the balance is maintained.

In the equation 7 is **added** to s on the LHS.
Subtracting 7 from both sides gives the value of s:

$$s + 7 - 7 = 35 - 7$$
i.e. $s \qquad = 28$

The **solution** is $s = 28$
This tells us that the shirt costs £28.

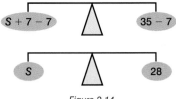

Figure 2.14

Adding and subtracting are **inverse** operations (opposites).
Equations can be solved using inverses.

 Whatever you do to one side, do the same to the other side.

Example

Ben is b years old. Four years ago, Ben was 15.
a Write down and solve an equation for b.
b How old will Ben be in two years' time?

Here's how...

a Four years ago Ben's age was 4 less than b (i.e. $b - 4$)
The equation is $b - 4 = 15$
To find the value of b, **add** 4 to both sides:

$$b - 4 = 15$$
$$b - 4 + 4 = 15 + 4$$
i.e. $$b = 19$$

Answer: Ben is 19

b Ben is 19 now. In two years' time, he will be 21 years old.

Answer: In two years, Ben will be 21 years old

hint

You can probably see that Ben must be 19. How did you get this? Compare what you did with the way the equation is solved later.

hint

The inverse of -4 is $+4$.

hint

After practice, you can shorten the working:

To solve $b - 4 = 15$
write $b = 15 + 4$
            $b = 19$

Equations can involve multiplying or dividing

Suppose five students decide to share a restaurant bill equally. The bill is £65.
How much does each student pay?
(Clearly, the answer is £65 ÷ 5 = £13. Compare this with the working in the equation below.)

The information can be written as an equation using c to stand for each student's contribution (in £).
The total contribution from 5 students is $5 \times c$, or simply $5c$. This must equal the total (65).

The equation is $5c = 65$

In the equation c is **multiplied** by 5. The solution is found by **dividing** both sides by 5:

$$\frac{\cancel{5}c}{\cancel{5}} = \frac{65}{5}$$

Cancelling the 5's gives $c = 13$

This confirms the students each pay £13 towards the bill.

Multiplying and dividing are another pair of **inverse** operations.

Adding and subtracting are inverse operations
Multiplying and dividing are inverse operations

Example

An unknown sum of money, £p, is shared equally between 6 people.
a Write down an expression for the amount in £ each person receives.
b If each person receives £4.50, write down and solve an equation for p.

Here's how...

a The amount received by each person is $\dfrac{£p}{6}$

b The equation is: $\dfrac{p}{6} = 4.50$

Multiply by 6: $p = 4.5 \times 6 = 27$

Answer: The sum of money was £27

hint

An 'expression' can include numbers and letters. An 'equation' has an equals sign (=) as well.

hint

In the equation p is *divided* by 6. To solve it, *multiply* by 6.

checkpoint

Check that the answer fits the original information.
£27 ÷ 6 = £4.50, correct!

> When you solve an equation, *always check that the solution fits the original information.*

Now try these...

1   **a**   A box contains n mints. Ian eats 9 mints. Write down an expression for the number of mints left.
     b   Ian counts 16 mints left in the box. Write down and solve an equation to find the value of n.

2   A milkman delivers either 'half-fat' or 'full-fat' milk. The milk float holds h 'half-fat' and 96 'full-fat' milk bottles. The total number of milk bottles on the float is 252. Write down and solve an equation for h.

3   A box contains l '40 watt' light bulbs. The total power of the bulbs in the box is 480 watts.
     a   Write down and solve an equation for l
    One-third of the bulbs are replaced by '60 watt' bulbs.
     b   What is the total power of the bulbs in the box now?

4   Anne has £m in her bank account. After spending £399 on a new stereo she has £27 left.
     a   Write down and solve an equation for m.
    Anne sells her old stereo for £$\frac{1}{2}m$ and pays the money into her account.
     b   What is the balance in her account?

5   A hiker walks a triangular route ABC around the outside of a lake. The route starts and finishes at point A. The hiker walks at a steady speed of 6 km/h. He takes $2\frac{3}{4}$ hours to complete the walk.

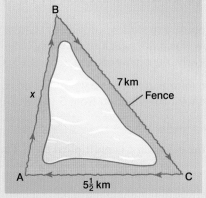

Figure 2.15

     a   Calculate the total distance the hiker walked.
     b   Write down and solve an equation for x.
    When the hiker returns to A, he realises he has lost his gloves. He thinks this happened when he took them off to climb a fence situated $\frac{3}{5}$ of the way from B to C.
     c   Is the shorter route around the triangle to the fence via point B or point C?
       Show working to support your answer.

6   Sam organises a day trip to London for himself and five friends. He buys 6 one-day travel cards. Each travel card costs £c. Sam pays for the travel cards with a £20 note and receives £4.40 change.
     a   Write down and solve an equation for c.
    During the trip Sam loses his wallet and travel card. Sam's friends each give him an equal amount of money to buy a replacement card.
     b   How much does each friend give to Sam?

7   Computer disks are sold in boxes of 8.
     a   Write down an expression for the number of disks in b boxes.
    In b boxes of disks, 248 disks work and 8 are faulty.
     b   Write down and solve an equation for b.
     c   All the faulty disks were in the same box. Calculate the percentage of boxes containing non-faulty disks.

8   Jack has a collection of 5 pence and 10 pence coins. He has twice as many 10 pence coins as 5 pence coins. The total value of the collection is £6.75.
     a   If x is the number of 5 pence coins in the collection, write down an expression for the number of 10 pence coins.
     b   Explain why $25x = 675$.
     c   By solving the equation from part **b**, find the total number of coins in the collection.
    Jack adds six 5 pence coins and three 10 pence coins to the collection.
     d   What is the ratio of 5 pence to 10 pence coins in the collection now?
       Choose the correct answer from: **A** $1:2$ **B** $11:19$ **C** $2:1$ **D** $15:17$

Equations can involve more than one operation

The equation $3g + 6 = 21$ involves both multiplication *and* addition.

There are two **terms** on the LHS, $3g$ and 6.

6 is called a **constant** term – it does not involve a letter.

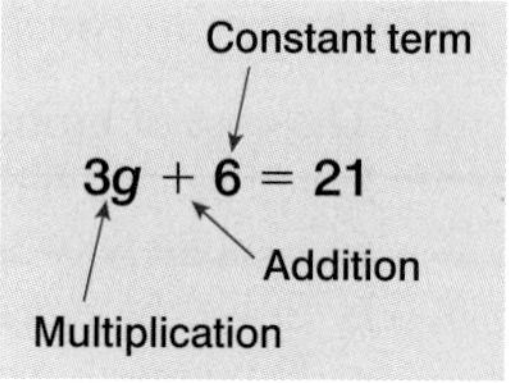

Constant term

$$3g + 6 = 21$$

Addition

Multiplication

Figure 2.16

Example

The owner of a Chinese restaurant orders boxes of noodles and bags of rice. One box of noodles costs £3. One bag of rice costs £p. An order of 5 boxes of noodles and 3 bags of rice costs £18.60.

a Write down and solve an equation for p.
Each bag of rice has mass 2 kg.
b What is the cost per kg of rice?

Here's how...

a Cost of the noodles $= 5 \times £3 = £15$. Cost of the rice $= £3p$.

The total cost gives the equation $\qquad 3p + 15 = 18.6$

Remove the constant term first ... $\qquad 3p = 18.6 - \mathbf{15}$

$$3p = 3.6$$

... then divide both sides by 3 ... $\qquad p = \dfrac{3.6}{3} = 1.2$

Answer: 1 bag of rice costs £1.20

b A 2 kg bag of rice costs £1.20. The cost per kg $= £1.20 \div 2$
$$= £0.60$$

Answer: Cost per kg = 60 pence

hint

$3 \times £p = £3p$

hint

To remove **+15** from the LHS, *subtract* 15.

checkpoint

Using the original information:
If a bag of rice costs £1.20, 3 bags of rice + 5 boxes of noodles cost
$3 \times £1.20 + 5 \times £3$
$= £3.60 + £15$
$= £18.60$, correct!

Example

Abdullah invests £V in a building society account.
When the investment has doubled in value he withdraws £50.

a Write down an expression for the amount left in the account.
b The investment is then worth £1000. Write down an equation for V.
c Find the size of the original investment.

Here's how...

a Abdullah invests £V. The investment doubles to give £$2V$.
£50 is withdrawn. The amount left is **£$(2V - 50)$**.

b The investment is then worth £1000. This gives the equation:
$$\mathbf{2V - 50 = 1000}$$

c $\qquad 2V = 1000 + \mathbf{50}$

$$2V = 1050$$

$$V = \frac{1050}{2} = 525$$

Answer: Investment was £525

hint

Brackets are used because all of the expression is in £.

checkpoint

Check the original information
$2 \times £525 - £50 = £1050 - £50$
$= £1000$
Correct!

Now try these...

Remember to check each answer.

1   The cost of hiring a motor boat includes a basic charge of £10 plus £p for each hour.
    When Angela hires the motor boat for 3 hours it costs her £28.
    a Explain why $3p + 10 = 28$
    b Solve the equation to find the value of p.
    c How much would it cost to hire the motor boat for 5 hours?

2   Neil has a part-time job. His rate of pay is £5 per hour.
    a Write down an expression for Neil's earnings on a day when he works t hours.
    b Neil spends £1 on bus fares and £2 on lunch. Explain why the amount he takes home is £$(5t - 3)$.
    c If Neil takes home £27, write down an equation and use it to find t. Interpret your solution.

3   An empty oil can weighs 200 g. Two litres of oil are poured into the can.
    a Using w grams to represent the weight of one litre of oil, write down an expression for the total
       weight of the oil can and its contents.
    b The total weight is 2 kg. Write down an equation and solve it to find the density of oil in grams per
       litre.

4   In a training session Mary completes 6 laps of a running circuit. She finishes the first lap in
    50 seconds. As she begins to tire, each lap after the first takes Mary t seconds more than the
    previous lap.
    a Explain why an expression for the time taken to run the *third* lap is given by $50 + 2t$.

    It takes her 80 seconds to run the final lap.
    b Write down and solve an equation for t.
    c Calculate Mary's total running time for the 6 laps.

5   £p is invested on the stock exchange. By the end of Year 1, the investment has doubled.
    a Write down an expression for the value of the investment at the end of Year 1.

    During Year 2, the investment trebles in value.
    b Write down an expression, in terms of p, for the value of the investment at the end of Year 2.

    During Year 3, there is a crash on the stock exchange. The investment loses £2000 in value.
    It is then worth £1450.
    c Write down and solve an equation for p.
    d Estimate the percentage increase in the investment over the three years.

6   Bob buys eight wooden posts, each 20 cm wide, to make a fence. The fence must fit a space
    10.7 m wide. Bob wants to find out how far apart the posts need to be placed.
    This distance, d metres, is shown on the sketch. Write down and solve an equation for d.

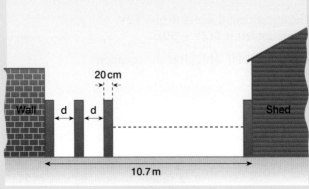

Figure 2.17

Remove brackets by multiplying out

Example

The formula $F = 10(P - 2)$ gives the number of French francs received in exchange for £P. Sue exchanges £P and receives 1980 francs.
a Write down and solve an equation for P
b How many francs per £ does Sue receive?

Here's how...

a The equation is: $\quad 10(P - 2) = 1980$
Remove the brackets: $\quad 10P - 20 = 1980$
$$10P = 1980 + \mathbf{20}$$
$$10P = 2000$$
$$P = \frac{2000}{\mathbf{10}} = 200$$

Answer: Sue exchanged £200

b Sue receives 1980 francs for £200.
Number of francs per £ $\quad = 1980 \div 200$
$$= 9.9 \ldots$$

Answer: 9.9 francs per £

hint

Both terms inside the bracket are multiplied by 10.

checkpoint

LHS $= 10(P - 2)$
$= 10(\mathbf{200} - 2)$
$= 10 \times 198$
$= 1980$

hint

$\dfrac{198\cancel{0}}{20\cancel{0}} = \dfrac{198}{20} = \dfrac{99}{10} = 9.9$

Now try these...

Remember to check each answer

1 The length of tape, t cm, needed to edge a square blanket is given by the formula $t = 4(l + 2.5)$ where l cm is the length of each side. The length of tape used to edge one particular blanket is 850 cm. Write down and solve an equation for l.

2 Each of Sally's training sessions in a gym lasts t minutes. She always spends the first 5 minutes setting up the equipment and the rest of the time training. In six sessions she has spent a total time of 4 hours training.
   a Explain why $6(t - 5) = 240$
   b Solve the equation to find the value of t.
   c If each session costs £7.50, how much will it cost Sally for a total training time of 10 hours?

3 The total mass of 6 cans of beans is 3 kg. Each empty can weighs 80 g.
   a If the mass of beans in each can is m g, explain why $6(m + 80) = 3000$
   b Solve the equation to find the value of m.
   c A total mass of 6 kg of beans is required for a meal at scout camp. How many cans of beans are needed?

4 A sum of money, £m, is invested in a company. An end of year bonus increases its value by £100. A take-over bid then doubles the value of the investment.
   a Write down an expression using brackets for the value of the investment after the take-over bid.
   After the take-over, £125 is deducted from the investment for commission. The value of the investment is now £1575.
   b Write down and solve an equation for m.
   c How many times greater is the investment now compared to the original sum?

2.2.4 Solving more difficult equations

LEVEL 3 **Collect terms**

If the equation has more than one term containing the unknown, collect these terms together on one side of the equation.

Example

The amount charged by two companies for printing n copies of a poster are:

Print-it Right   Charge in £ $= 0.15n + 4.8$
Copyprint   Charge in £ $= 0.2n + 3.5$

Find the number of copies for which both companies charge the same.

Here's how...

The equation is: $\qquad\qquad 0.2n + 3.5 = 0.15n + 4.8$

Subtract 0.15n: $\quad 0.2n + 3.5 - \mathbf{0.15n} = 4.8$

$\qquad\qquad\qquad\qquad 0.05n + 3.5 = 4.8$

Subtract 3.5 $\qquad\qquad\qquad 0.05n = 4.8 - \mathbf{3.5}$

$\qquad\qquad\qquad\qquad\qquad 0.05n = 1.3$

$$n = \frac{1.3}{\mathbf{0.05}} = 26$$

Answer: The companies charge the same amount for 26 posters

hint

Print-It Right charges a basic fee of £4.80 plus 15p per copy. Copyprint charges a basic fee of £3.50 plus 20p per copy.

hint

Subtract 0.15n to collect the terms in n on the LHS.

hint

Subtract the number of n's: $0.2n - 0.15n = 0.05n$

checkpoint

Print-it Right charges $0.15 \times 26 + 4.8 = 8.7$ (£) Copyprint charges $0.2 \times 26 + 3.5 = 8.7$ (£) Correct!

Now try these...

Remember to check each answer

1 Plumber A charges a call-out fee of £10 plus £20 for each hour he spends on the job. Plumber B charges a call-out fee of £20 plus £16 per hour. Their charges are the same for a job lasting h hours.
   a Explain why $10 + 20h = 20 + 16h$.
   b Solve the equation to find the value of h.

2 Mary says 'Twelve years from now I will be three times as old as I am now'.
   a Show that $3m = m + 12$ where m is Mary's present age in years.
   b Solve the equation to find how old Mary is now.

3 Sue, Glyn and Dan deliver newspapers. One week Sue earns £3 more than Dan whilst Glyn earns twice as much as Sue.
   a If Dan earns £x write down an expression for the amount earned by   **i** Sue   **ii** Glyn.
   b If altogether they earn £55, write down and solve an equation for x.
   c Find the amount earned by each person.

4 Pete and Jim compete in a cycling race. Jim cycles at 15 metres per second and finishes the race 20 seconds before Pete who cycles at 14.4 metres per second.
   a If Pete takes t seconds to complete the race, explain why $15(t - 20) = 14.4t$.
   b Find the time taken by each cyclist to complete the race.
   c Find the length of the race, giving your answer in kilometres.

LEVEL 3
The inverse of squaring is taking the square root

For example, $4^2 = 16$ and $\sqrt{16} = 4$. (NB -4 is also a square root of 16, but in real-life problems negative values are not usually needed.) To remove a cube, take the cube root. For example, if $x^3 = 125$, then $x = \sqrt[3]{125} = 5$. A calculator can be used to find more difficult roots (See ToTT. 7).

Example

A plot of land consists of three identical squares (as shown in the diagram). The total area of the land is 150 m².
A gardener wishes to surround the outside of the land with a fence. Calculate the length of fencing required.
Give your answer to an appropriate degree of accuracy.

Figure 2.18

hint

A square with side l has area $l \times l = l^2$. Perimeters and areas are studied more fully in Section 2.3.

Here's how...

Let the side of each square be l metres.
First we must find l.
The total area is $\quad l^2 + l^2 + l^2 = 3l^2$
Equation: $\qquad\qquad 3l^2 = 150$
$\qquad\qquad\qquad l^2 = 50$
$\qquad\qquad\qquad l = \sqrt{50}$
$\qquad\qquad\qquad\quad = 7.071$ m

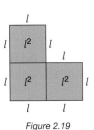

Figure 2.19

hint

Ignore the negative square root.

Length of fencing required is $8 \times 7.071 = 56.568$ m

Answer: The gardener should buy 60 m of fence

hint

It is sensible to round up, allowing extra fencing.

Now try these...

Remember to check each answer.

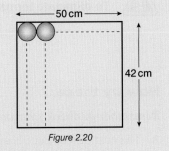

Figure 2.20

1 The formula $A = \pi r^2$ gives the area A of a circle (in cm²) with radius r (cm). A circular place mat has area 78.5 cm².
   a Write down and solve an equation for r. Give your answer to the nearest whole cm.
   A rectangular coffee table measures 50 cm by 42 cm, as shown.
   b What is the largest number of place mats that can be placed side by side on the table?

2 The volume of a sugar cube is given by the formula $V = x^3$, where x is the length of each side. Find the length of the side of a sugar cube whose volume is 900 mm³.

3 A building society account gives a fixed rate of interest, $r\%$. If $£P$ is left in this account:
   a explain why the amount in the account at the end of n years is given by the formula $A = P\left(1 + \dfrac{r}{100}\right)^n$
   b find, to one decimal place, the rate of interest needed for £200 to grow to £240 in 3 years.

LEVEL 3
2.2.5 Finding two unknowns

If there are two unknowns to find, two equations are needed. Two equations involving two unknowns are called a pair of **simultaneous equations**.

Suppose Mr and Mrs Brown and their two children pay a total of £10 for a fairground ride and Mr and Mrs Jones and their young son pay a total of £8 for the same ride. This information can be used to find the cost of the ride for one child and the cost for one adult.

The table below shows how the problem can be solved by reasoning and also by using equations.

In the equations a represents the cost (in £) for an adult, and c represents the cost (in £) for a child. The equations are numbered so that each step in the working can be explained.

Reasoning:	Using equations:
Browns: 2 adults and 2 children pay £10	$2a + 2c = 10$ **(1)**
Jones: 2 adults and 1 child pay £8	$2a + c = 8$ **(2)**
The extra £2 must be the cost for a child	**(1) – (2)** $c = 2$
Jones: Cost for 2 adults must be £8 – £2 = £6	Using **(2)** $2a + 2 = 8$
	Subtract 2 $2a = 6$
Cost for 1 adult must be £6 ÷ 2 = £3	Divide by 2 $a = 3$
Answer: Costs are £2 for a child, £3 for an adult	**Costs are £2 for a child, £3 for an adult**
Check 2 adults and 2 children pay:	**Check** in **(1)**
Browns: £6 + £4 = £10 Correct!	$2a + 2c = 6 + 4 = 10$ Correct!

NB Subtracting the equations results in the disappearance of letter a. Both of the $2a$ terms disappear. This happens because $2a - 2a = 0$. The unknown value a is **eliminated**.

Sometimes it is necessary to add the two equations in order to eliminate one of the unknowns. For example, in the simultaneous equations: $\qquad 3p + 2q = 19 \qquad$ **(1)**
$\qquad\qquad\qquad\qquad\qquad\qquad\qquad\qquad\qquad 5p - 2q = 5 \qquad$ **(2)**
the 'q' terms have different signs. Adding **(1)** and **(2)** eliminates the 'q' terms, because $2q + -2q = 0$

> To eliminate terms with the *same* signs, *subtract* the equations.
> To eliminate terms with *different* signs, *add* the equations.

Now try these...

Remember to check your answers

1 The cost of four pencils and an eraser is £1. The cost of two pencils and an eraser is 70 pence.
   a If p pence is the cost of a pencil and r pence is the cost of an eraser, explain why $4p + r = 100$.
   b Write down another equation relating p and r.
   c By solving the equations find the price of a pencil and the price of an eraser.

2 A suit consists of a jacket costing £j and a pair of trousers costing £t. The suit costs £111. The jacket costs £39 more than the trousers.
   a What is the value of: **i** $j + t$ **ii** $j - t$?   (Write each answer as an equation.)
   b Solve the equations found in part **a**.
   A week later, the same suit is on sale. The cost of the jacket is reduced by 20% and the trousers by 25%.
   c What is the sale price of the suit?

3 A catering firm has five teapots, each with capacity t litres, and four coffee pots, each with capacity c litres. The total capacity of the teapots is the same as the total capacity of the coffee pots. When these totals are added together the overall capacity is 12 litres.
   a Explain why $5t - 4c = 0$ and find another equation relating t and c.
   b Solve the equations to find the capacity of a teapot and the capacity of a coffee pot.

4 Sam works a basic 16 hour week for a newsagent. His basic rate of pay is £w per hour. His overtime rate is £t per hour. One week, Sam does two hours overtime and is paid a *total* of £104.50.

   a Write down an equation relating w and t.

In the next week, Sam does five hours overtime and earns a total of £129.25.

   b Write down a second equation relating w and t.

   c Solve the pair of simultaneous equations.

   d Express the overtime rate in terms of the basic rate – e.g. double time, time and a half, etc …

Sam's pay is reviewed. He is awarded an extra £1.50 per hour on his basic rate. Overtime is paid at twice the basic rate (i.e. 'Double Time')

   e How much does Sam earn in total in a week that includes 3 hours overtime?

More difficult equations may need multiplying first

Sometimes neither of the unknowns can be eliminated immediately because the terms containing them in the equations are different. In this case, multiply one or both of the equations to give other equations that do contain a pair of equal terms.

For example if the equations were:

$$4u + 3v = 18 \ldots \textbf{(1)}$$
$$5u - 2v = 11 \ldots \textbf{(2)}$$

adding equations **(1)** and **(2)** would *not* eliminate the 'v' terms.

Instead, a new pair of equations must be formed as follows:

Multiply each number in **(1)** by 2: $8u + 6v = 36 \ldots \textbf{(3)}$
Multiply each number in **(2)** by 3: $15u - 6v = 33 \ldots \textbf{(4)}$

Adding equations (3) and (4) would now eliminate v and allow u to be found.

The next example shows the method in full.

> *hint*
>
> because $3v + -2v = v$

> *hint*
>
> **NB** The multiplier for **(1)** is the number in front of v in equation **(2)**. The multiplier for **(2)** is the number in front of v in equation **(1)**.

Example

Admission to a leisure centre costs £20 for a party of two adults and five children. A group of four adults with three children pay an extra £4.60. Find the admission charge for one adult with two children.

Here's how...

Let a be the charge for one adult and c the charge for one child (both in £).

Write the information as a pair of equations:	$2a + 5c = 20$	**(1)**
	$4a + 3c = 24.6$	**(2)**
Multiply **(1)** by 3:	$6a + 15c = 60$	**(3)**
Multiply **(2)** by 5:	$20a + 15c = 123$	**(4)**
Subtract **(4)** − **(3)** to eliminate c	$14a = 63$	
Divide by 14 to find a	$a = \mathbf{4.5}$	
Substitute 4.5 for a in **(1)**	$2 \times 4.5 + 5c = 20$	
	$9 + 5c = 20$	
Subtract 9	$5c = 11$	
Divide by 5	$c = \mathbf{2.2}$	

Admission charges are £4.50 for an adult and £2.20 for a child.

Cost for 1 adult and 2 children   = £4.50 + 2 × £2.20
                                   = £4.50 + £4.40

                                         Answer = £8.90

> *hint*
>
> To make the c terms equal: multiply **(1)** by the number in front of c in **(2)**; multiply **(2)** by the number in front of c in **(1)**.

> *hint*
>
> The numbers are slightly easier in **(1)** than **(2)**.

> *checkpoint*
>
> Cost for 2 adults and 5 children:
> = 2 × £4.50 + 5 × £2.20
> = £9 + £11 = £20
> Cost for 4 adults and 3 children:
> = 4 × £4.50 + 3 × £2.20
> = £18 + £6.60 = £24.60
> i.e. £4.60 more, correct!

> *hint*
>
> Make sure you answer the question!

Summary of method

Step 1 Multiply one, or both, equations to *make the terms in one unknown equal* (if necessary).

Step 2 Look at the signs of the equal terms. If the *signs are the same, subtract* the equations. If the *signs are different, add the equations*.
This should give you an equation involving just one of the unknowns.

Step 3 *Solve the equation to find the value of one unknown.*

Step 4 *Substitute* the value you have found into one of the original equations to find the second unknown.

Step 5 *Check* both values using the original information.

Now try these...

Remember to check your answers

1 Let b be the cost (in £) of one pint of beer and c the cost (in £) of one packet of crisps.
   a Two pints of beer and three packets of crisps cost a total of £4. Write down an equation in b and c.
   b Three pints of the same beer and four packets of the same crisps cost a total of £5.86.
   Write down a second equation in b and c.
   c Solve the equations and interpret your solutions.
   d What is the total cost of a round of four pints of beer and four packets of crisps?

2 A collection of 26 coins consists of 5 pence and 2 pence pieces. The collection has a total value of £1. Let f be the number of 5 pence pieces and t the number of 2 pence pieces.
   a Explain why $5f + 2t = 100$
   b Find a second equation in f and t.
   c Solve the equations and interpret your solutions.
   d How many times greater in total value are the 5 pence pieces than the 2 pence pieces?

3 A bank account is opened. Six equal deposits of £d are made followed by four equal withdrawals of £w. The balance in the account is then £130.
   a Write down an equation in d and w.
   A further five equal deposits of £d and three equal withdrawals £w leaves a balance of £245.
   b Explain why $5d - 3w = 115$
   c Solve the equations.
   d How many more withdrawals of £w can be made before the account becomes overdrawn?

4 A small theatre has two types of seats: stall seats and circle seats. A stall seat costs £s and a circle seat £c. The table shows how many seats were bought and the total takings on two consecutive days.

	No. stall seats sold	No. circle seats sold	Total takings
Day 1	52	29	£1065
Day 2	48	21	£885

Table 2.34

   a Explain why $52s + 29c = 1065$
   b Write down a second equation.
   c Solve the equations.

On the third day, the manager sells 50 stall seats and 25 circle seats.
   d Find the total takings from seats on the third day.
   e How can *adding* equations from **a** and **b** help you check your answer to part **d**?

5 A door to door salesman buys b hairdryers. In one day, he sells s of the hairdryers. At the end of the day, he has 4 hairdryers left.
  a What is the value of $s - b$? {take care with the signs!}
  The salesman bought the hairdryers for £5 each and sold them for £9.50 each. He makes a profit of £52.
  b Write down another equation in s and b.
  c Solve the equations.
  The salesman sells the remaining hairdryers at £7.50 each.
  d Calculate: **i** the total profit made
                     ii the average profit made per hairdryer ($=$ Total profit $\div$ No. hairdryers bought)

LEVEL 3 2.2.6 Rearranging formulae

Amy is three years older than Ben. If a stands for Amy's age and b stands for Ben's age then a formula for Amy's age in terms of Ben's age is: $a = b + 3$
                                                                                                                                                      $\uparrow$
                                                                                                                                                   subject

The **subject** is the letter that stands alone at the left hand side of the equals sign.

If we want Ben's age in terms of Amy's age, then the formula must be **rearranged** to make b the subject.
To rearrange the formula, use inverse operations i.e. do the opposite to both sides.
It helps to turn the formula around first: $b + 3 = a$ (b is then on LHS)
Subtract 3, to leave b on its own: $b = a - 3$
The rearranged formula makes sense: $\uparrow$
(Ben is 3 years younger than Amy)   new subject

Example
The formula $R = \dfrac{V}{I}$ gives the relationship between the resistance (R ohms),
the voltage (V volts) and the current (I amps) in an electrical circuit.
  a Rearrange the formula to make the subject: **i** V **ii** I
  b Find the voltage when the current is 5 amps and the resistance is 12 ohms.

Here's how...
  a i Turn around the given formula: $\dfrac{V}{I} = R$

     Multiply by I to leave V on its own: $V = IR$

                                                   Answer: $V = IR$

> *hint*
> V is *divided* by I, so *multiply* by I.

   ii Turning around the answer for part **a**: $IR = V$

     Divide by R to leave I on its own: $I = \dfrac{V}{R}$ **Answer: $I = \dfrac{V}{R}$**

> *hint*
> To find V use the formula that has V as the subject.

  b Substitute $I = 5$ and $R = 12$ in $V = IR$
     $V = 5 \times 12 = 60$

                   Answer: The voltage is 60 volts.

Note that it is possible to divide by more than one letter in a single step.

For example, dividing $mgh = P$ by both m and g gives $h = \dfrac{P}{mg}$

It is also possible to add or subtract a complex term in a single step. For example, in the formula $E - mgh = K$, adding the term mgh gives $E = K + mgh$. (These are all formulae involving energy.)

Now try these...

1 The total mass of a box and its contents is given by the formula $T = B + C$ where T is the total mass, B is the mass of the box and C is the mass of the contents (in kilograms).
   Make C the subject of the formula.

2 The formula $P = \dfrac{F}{A}$ gives the pressure, P, on a surface. F is the total force and A is the surface area.
   Make the subject of the formula: **a** F **b** A

3 Energy in food can be measured in calories or in joules. The formula $J = 4.2C$ gives the approximate conversion for kilocalories into kilojoules.
   a Rearrange the formula to make C the subject.
   b The energy per 100 g in a tin of cooking sauce is 334 kilojoules. The tin holds 390 g of sauce. Estimate the energy level of the sauce in kilocalories.

4 Gerald rearranges the formula $v = u + at$ to make u the subject. His answer is $u = \dfrac{v}{at}$
   a What mistake has Gerald made?   **b** Write down the correct answer.

5 The formula $P = 2\pi r$ gives the perimeter, P metres, of a circular pond whose radius is r metres.
   a Make r the subject of the formula.   **b** Find the radius of a pond if its perimeter is 6 metres.

Rearranging a formula may need several steps

Aim to leave the required subject alone at one side.
Remove the other terms using inverses, but **don't try to do everything at once** – just do one thing at a time.

Example

The formula $A = 16h + 12$ gives the amount charged A (£) by an electrician working for h hours.
a Rearrange the formula to make h the subject.
b How long did the electrician work on a job for which he charged £64?

Here's how...

a Given formula: $16h + 12 = A$

   Subtract 12: $16h = A - 12$

   Divide by 16 $h = \dfrac{A - 12}{16}$

   Answer: $h = \dfrac{A - 12}{16}$

> *hint*
> Remove the constant term, 12, first.

b Substitute $A = 64$ into the rearranged formula for h:

   $h = \dfrac{A - 12}{16} = \dfrac{64 - 12}{16} = 3.25$

   Answer: The job took $3\frac{1}{4}$ hours

> *hint*
> $\dfrac{64 - 12}{16} = \dfrac{52}{16} = \dfrac{13}{4}$

If a formula has a square term then rearranging may need the use of square roots.

Example

A stone is released from the top of a building. The distance d (metres) travelled by the stone in the first t seconds of its fall is given by $d = \frac{1}{2}gt^2$ where g is the acceleration due to gravity (in m/s²).
a Rearrange this formula to make t the subject.
   The building is 24 metres tall and g is approximately 9.8 (in m/s²).
b Calculate the time taken for the stone to hit the ground.

> *hint*
> m/s² means 'metres per second per second'. It gets faster by 10 m/s every second.

Here's how...

a Given formula: $\frac{1}{2}gt^2 = d$

Multiply by 2 $\quad gt^2 = 2d$

Divide by g $\quad t^2 = \dfrac{2d}{g}$

Take the square root $\quad t = \sqrt{\dfrac{2d}{g}}$ **Answer:** $t = \sqrt{\dfrac{2d}{g}}$

hint

The opposite of 'halving' is 'doubling'. To remove $\frac{1}{2}$, multiply by 2.

hint

t^2 must be on its own before the square root is taken.

b Substitute $d = 24$ and $g = 9.8$ into the formula for t.

$$t = \sqrt{\frac{2 \times 24}{9.8}} = \sqrt{4.8979} = 2.2131\ldots$$

Answer: $t = 2.2$ seconds (2 s.f.)

hint

Assuming 24 and 10 are correct to 2 s.f., round the answer to 2 s.f.

Now try these...

1 The cost £C of hiring a van for d days is given by $\quad C = 15d + 20$
   a Rearrange the formula to make d the subject.
   b How long was a van kept if the cost was £110?

2 The area of a triangle is given by $A = \dfrac{bh}{2}$ where b is the length of the base and h is the height.

Sarah correctly rearranges the formula to make the subject h. Choose Sarah's correct answer from:

A $h = 2A - b$ **B** $h = \dfrac{b}{2A}$ **C** $h = \dfrac{2A}{b}$ **D** $b = \dfrac{2A}{h}$ **E** $h = 2(A - b)$

3 A clock strikes 12 noon. The formula $L = \dfrac{\pi DT}{60}$ gives the distance L cm travelled by the top of the

minute hand during a period of T minutes. D cm is the diameter of the clock.
   a Rearrange the formula to make D the subject.
   By 12:33 the tip of the minute hand has travelled a distance of 19 cm.
   b Calculate the diameter of the clock face (to the nearest cm).
   c How far has the minute hand travelled by 2:45 pm the same afternoon?

4 The number of people queuing at a bus stop is recorded every minute. The numbers fit the formula:
$N = 4T + 2$ where N is the number of people at the bus stop T minutes after the previous bus has left.
   a Use the formula to find the number of people queuing after 4 minutes.
   b Rearrange the formula to make T the subject.
   c After how many minutes will 25 people be queuing? (Use an appropriate level of accuracy.)
   After 6 minutes there are only 4 people recorded in the queue.
   d Give one explanation as to why the answer to part **c** does not agree with the recorded number.

5 The formula $M = \dfrac{2.2T}{N}$ gives the average mass M (in lb) of N people. T is the total mass of people in kg.

   a Rearrange the formula to make T the subject.
   b A group of 4 people have an average mass of 165 lb. Calculate the total weight of the group in kg.
   c The heaviest person weighs 79.2 kg. The lightest person weighs 71.6 kg. The other two people
   have equal mass. Find this mass.

6 The total cost C (in £) of hiring a car for D days is given by $C = 25D + 0.15M$ where M is the total
number of miles driven in the car.
   a Rearrange the formula to make M the subject.
   b Carol is on holiday in the Lake District. She hires a car for 3 days and is charged £123.
   How many miles has Carol driven?
   c When Carol returns the car, the mileometer reads 8405 miles. John is the next person to use the
   car. He hires the car for 4 days and is charged £162.25. What is the reading on the mileometer
   when John returns the car?

7   Oil spilled from a tanker forms a circular slick on the sea. The
formula $V = \pi dr^2$ gives the volume V (in m³) of the slick when
its depth is d metres and its radius is r metres. As the slick
spreads, its radius increases and its depth decreases.
   a Rearrange the formula to make r the subject.
   b The total volume V of the oil spilled by the tanker is
       700 m³. At 12 noon the slick has depth $d = 12$ cm on the
       sea's surface. Calculate the radius of the slick (to the
       nearest metre) at 12 noon.
   c By 2 pm the slick has depth $d = 5$ cm.
       Calculate, to the nearest metre, the radius of the slick at 2 pm.
   d What is the rate of increase in the radius in metres per hour?
   e Assume the radius has been increasing at this rate since
       the start of the slick.
       Estimate the time at which the tanker spilled its load.

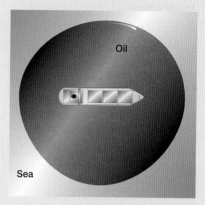

Figure 2.21

8   A stone is released from the top of a building. The formula $v = \sqrt{19.6d}$ gives the speed, v metres per
   seconds, of the stone after it has fallen d metres.
   a Find the speed of the stone after it has fallen 10 metres.
   b Rearrange the formula to make d the subject.
   c When it hits the ground, the stone has speed $v = 21$ m/s. Use your answer to part **b** to find the
       height of the building.

9   A hiker is walking roughly in a north-easterly direction (as shown).
   The formula $D = \sqrt{N^2 + E^2}$ gives the actual distance D (in km)
   walked. N km is the distance travelled due north. E km is the
   distance travelled due east.
   a Rearrange the formula to make N the subject.
   By using a map, the hiker knows he has actually walked 29 km. He
   is 21 km further east than when he started.
   b How many kilometres further north is he than when he started?
   c Which side of 'north-east' has the hiker been heading?

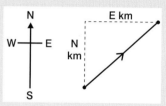

Figure 2.22

Rearranging formulae with brackets depends on the position of the required subject

In the formula: $\qquad$ $w = h(t - 4)$ $\qquad$ the letter h is *outside* the bracket.
Turning the formula around: $h(t - 4) = w$

To make h the subject, divide both sides by $(t - 4)$ giving $h = \dfrac{w}{(t - 4)}$

If the required subject lies *inside* a bracket, you can start by removing the bracket.

Example

The formula $p = 2(l + w)$ gives the perimeter p of a rectangle of length l and
width w. Rearrange the formula to make l the subject.

Here's how...
The required subject l is *inside* a bracket.

Given formula:	$2(l + w) = p$
Remove brackets:	$2l + 2w = p$
Subtract $2w$:	$2l = p - 2w$
Divide by 2:	$l = \dfrac{p - 2w}{2}$

Answer: $l = \dfrac{p - 2w}{2}$

hint

Alternative method:
dividing both sides by 2,
$$l + w = \frac{p}{2}$$
gives $l = \dfrac{p}{2} - w$
This looks different from
the other solution, but is
equally correct.

The alternative method shown above is particularly useful if the number outside a bracket is a fraction. **Remember that the opposite of halving is doubling.** To remove $\frac{1}{2}$, simply multiply both sides by 2.

Example

The formula $M = \frac{1}{2}(a + b)$ gives the average M of two numbers a and b.
Rearrange the formula to make a the subject.

Here's how...

Given formula: $\quad \frac{1}{2}(a + b) = M$

Multiply by 2: $\quad\quad a + b = 2M$

Subtract b: $\quad\quad\quad a = 2M - \boldsymbol{b}$ $\quad\quad\quad$ **Answer:** $a = 2M - b$

hint

Removing the bracket first gives the more complicated answer
$a = 2(M - \frac{1}{2}b)$

Now try these...

1 The formula $W = 6(b + 2t)$ gives Amardeep's total weekly wage W (in £). b is the basic number of hours she works per week. t is the number of hours overtime worked in the week.
   a Rearrange the formula to make b the subject.
   One week, Amardeep's total wage is £246. She worked 3 hours overtime.
   b How many hours is Amardeep's basic week?

2 The formula for converting temperatures from Fahrenheit to Centigrade is $C = \dfrac{5(F - 32)}{9}$ where F is the temperature in degrees Fahrenheit and C is the temperature in degrees Centigrade. Rearrange this formula to make the subject F.

3 The formula $I = P(1 + \frac{R}{100})^2$ gives the value I (in £) of an investment after 2 years. P is the original value of the investment and $R\%$ is the yearly interest rate. John invests £5000 at an annual interest rate of 4%.
   a Find the value of John's investment after two years.
   b Rearrange the formula to make P the subject.
   A certain investment grows at 3% per year. After two years, the investment is worth £7426.30.
   c Use your answer to part **b** to find the original value of the investment.

4 The exchange rate for converting UK pounds into Australian dollars is £1 : $2.5
   The exchange centre deducts £C in commission before the exchange is made.
   The formula $A = 2.5(P - C)$ gives the amount A (in $) received for exchanging £P.
   a Rearrange the formula to make P the subject.
   Trish and Ben want to travel to Australia. They have each saved some money in the ratio Trish:Ben = 3:4. They combine their savings and exchange the total for $1495 Australian dollars. The commission is £11.
   b How much (in £) did Trish and Ben exchange altogether?
   c How much more (in £) had Ben saved than Trish?

5 The formula $d = \dfrac{t(u + v)}{2}$ gives the distance, d metres, travelled by a car when it accelerates from u metres per second to v metres per second in t seconds. Make the subject $\quad$ **a** t $\quad\quad$ **b** v

6 Jill wants to buy a house worth £50 000. A fixed interest rate and term is agreed with a building society. Her mortgage repayments will depend on the amount Jill puts down as a deposit.
   The formula $R = \dfrac{7.24}{1000}(50\,000 - D)$ gives Jill's monthly repayment R (in £) for a deposit of £D.
   Jill would like to decide what she can afford each month and then work out what her deposit should be.
   a Show that $D = 50\,000 - \dfrac{1000R}{7.24}$ Explain each step in your working.
   Jill's net yearly income is £14 400. She wants her monthly repayment to be no greater than 25% of her net monthly income.
   b Calculate the greatest monthly repayment Jill wants to pay.
   c Calculate the minimum deposit Jill must put down if she is to afford the repayments.

Shape and Space

2.3.1 Working with plane shapes

Plane shapes are flat

A *plane shape* is a flat figure which forms a complete (or closed) shape.
Examples:

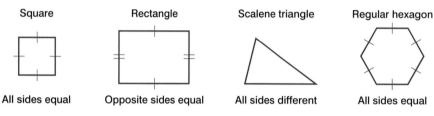

Square	Rectangle	Scalene triangle	Regular hexagon
All sides equal	Opposite sides equal	All sides different	All sides equal

Figure 2.23

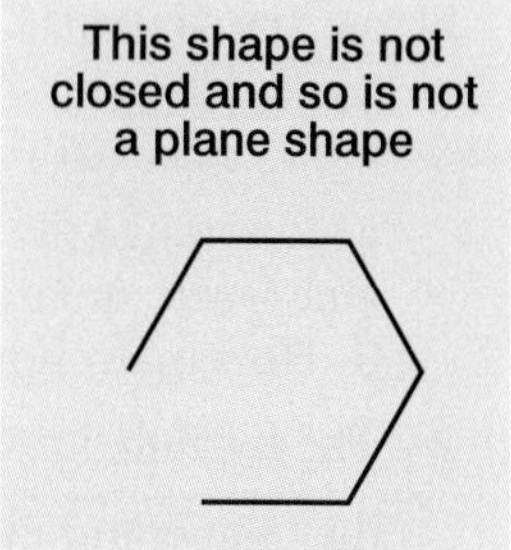

This shape is not
closed and so is not
a plane shape

Figure 2.24

The perimeter is the total length of all the outside edges

Example

A rectangular corridor has length 12.5 m and width 2.3 m (to 1 d.p.).
Find its perimeter.

Here's how...

The diagram shows the length of each side
of the corridor. Add all the lengths together.

$$\text{Perimeter} = 12.5 + 2.3 + 12.5 + 2.3$$
$$= 29.6 \text{ m}$$

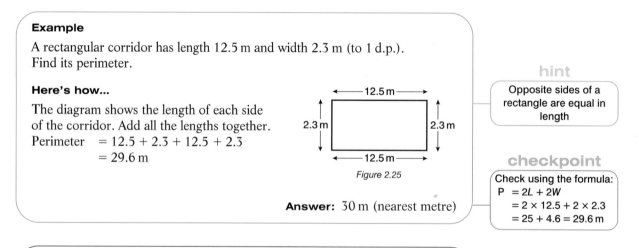

Figure 2.25

Answer: 30 m (nearest metre)

hint

Opposite sides of a
rectangle are equal in
length

checkpoint

Check using the formula:
$P = 2L + 2W$
$= 2 \times 12.5 + 2 \times 2.3$
$= 25 + 4.6 = 29.6 \text{ m}$

 A rectangle with length *L* and width *W* has
perimeter $P = 2L + 2W$

All measurements are approximate. The lengths given in the last example were correct to 1 decimal place, but
they may actually be slightly more or less than 12.5 m and 2.3 m. As a result the perimeter might be nearer to
29.5 m or 29.7 m than 29.6 m. Usually **answers to questions involving measurements are given to the same**
degree of accuracy as the *least* accurate measurement used, but it is often sensible to round even further. In
this case the answer was rounded to the nearest whole number of metres.

If some sides are given in different units, convert all lengths into the same units.

Example

A plan for a garden patio is shown.
(The dotted lines are not part of the patio).

Find the perimeter of the patio.

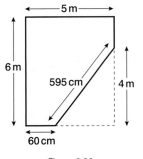

Figure 2.26

hint

For metric units see
Section 2.1.6.

Here's how...

Find any missing sides:

Most of the measurements are in metres.

Convert the lengths given in
centimetres into metres.

$595 \text{ cm} = 595 \div 100$ $60 \text{ cm} = 60 \div 100$
$\quad\quad\quad = 5.95 \text{ m}$ $\quad\quad\quad = 0.6 \text{ m}$

Perimeter $= 6 + 5 + 2 + 5.95 + 0.6$
$\quad\quad\quad = 19.55 \text{ m}$

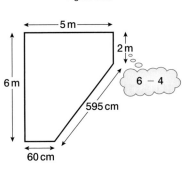

Figure 2.27

hint

The missing side
$= 6 - 4 = 2(m)$

checkpoint

Check by rounding:
$6 + 5 + 2 + 6 + 1 = 20$

hint

The accuracy of the
original measurements
was not given. It is
sensible to round the
answer to 19.6 m or 20 m.

Answer: 19.6 m (to 1 decimal place)

Now try these...

1 A garden pond is rectangular with sides of length 2.5 m and 3.4 m. A low wall is built around the edge of the pond. Find the length of this wall.

2 A football pitch is 120 m long and 90 m wide.
Find its perimeter.

3 The diagram shows the plan of an industrial site.
A security guard patrols the boundary of the site.
   a How far does the security guard walk each time
   he patrols the boundary?
   b Approximately how many times does the guard
   need to go around the boundary to walk 10 km?

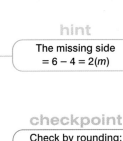

Figure 2.28

4 The diagram shows the plan of a field. The farmer
surrounds the field with a fence, leaving a
3 m gap for a gate.
What length of fence does the farmer use?

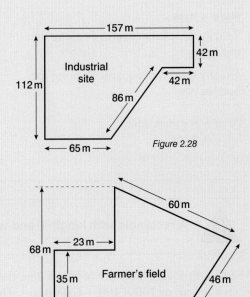

Figure 2.29

5 A rectangular rush mat of length 4.5 m and width 75 cm lies along a corridor. The edges of the mat are strengthened with tape. What length of tape is needed?

6 A manufacturer makes rectangular blankets that are 2.2 m long and 1.9 m wide. The blankets are edged with ribbon. The length of ribbon allowed for each blanket is equal to the perimeter of the blanket plus an extra 30 cm (in total) for turning corners and overlap.
Find the length of ribbon used to edge a batch of 200 blankets.

LEVEL 3

7 The perimeter of a rectangular flowerbed is 8.4 m. If the shorter sides of the flowerbed are 1.8 m long, what is the length of the longer sides?

8 The floor of a bandstand is in the shape of a regular hexagon with perimeter 15 m.
What is the length of each side?

Area is measured in square units

If the unit used is the 'cm' then the unit square is:

Figure 2.30(a)

If the unit is the 'inch' then the unit square is:

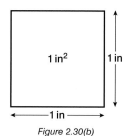

Figure 2.30(b)

The **area** of a plane shape is the number of unit squares that can fit inside the shape. The area depends on the unit of measurement used. An area measured in square inches will be numerically less than the answer in square centimetres because a square inch occupies more space than a square centimetre.

Example

A rectangular postage stamp has length 3 cm and width 2 cm.
Find the area of the stamp in square centimetres.

Here's how...

Exactly three unit squares fit along the top row.

Exactly three unit squares fit along the bottom row.

Area = 6 square units
    = 6 cm²

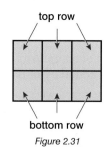

Figure 2.31

Answer: Area = 6 cm²

> **hint**
> Alternatively, use the formula:
> Area $A = L \times W$
> $= 3\,\text{cm} \times 2\,\text{cm}$
> $= 6\,\text{cm}^2$

 A rectangle with length *L* and width *W* has area $A = L \times W$

$1\,\text{m} = 100\,\text{cm}$

When a unit square has sides of length 1 m, the sides are also 100 cm long.
Its area is 1 m² or $100 \times 100 = 10\,000\,\text{cm}^2$. This gives a conversion factor for areas.
To convert m² to cm² multiply by 10 000. To convert cm² to m² divide by 10 000.

Round answers when necessary. Use the same number of **significant figures** (see ToTT.8) as in the **least accurate** of the measurements given. Round even further if it seems sensible to do so.

You may need to split up a complicated figure into basic shapes to find the area.

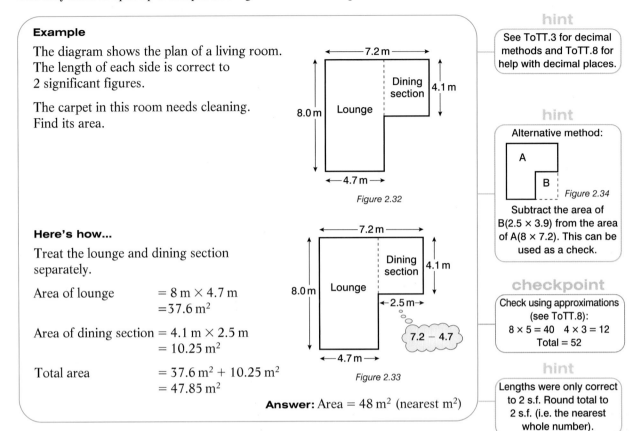

Example

The diagram shows the plan of a living room.
The length of each side is correct to
2 significant figures.

The carpet in this room needs cleaning.
Find its area.

Figure 2.32

hint

See ToTT.3 for decimal methods and ToTT.8 for help with decimal places.

hint

Alternative method:

Figure 2.34

Subtract the area of B(2.5×3.9) from the area of A(8×7.2). This can be used as a check.

checkpoint

Check using approximations (see ToTT.8):

$8 \times 5 = 40$ $4 \times 3 = 12$
Total = 52

hint

Lengths were only correct to 2 s.f. Round total to 2 s.f. (i.e. the nearest whole number).

Here's how...

Treat the lounge and dining section separately.

Area of lounge $= 8\text{ m} \times 4.7\text{ m}$
$= 37.6\text{ m}^2$

Area of dining section $= 4.1\text{ m} \times 2.5\text{ m}$
$= 10.25\text{ m}^2$

Total area $= 37.6\text{ m}^2 + 10.25\text{ m}^2$
$= 47.85\text{ m}^2$

Figure 2.33

Answer: Area $= 48\text{ m}^2$ (nearest m^2)

Now try these...

1   A manufacturer makes rectangular rugs in three different sizes. The length and width of each size are given below. Find the area of each size of rug.
    a Length 1.5 m, Width 1 m   **b** Length 2 m, Width 1.5 m   **c** Length 2.5 m, Width 1.5 m

2   The lengths of the sides of a passport photograph are 4.5 cm and 3.6 cm. Find its area.

3   A rectangular postage stamp is $1\frac{1}{2}$ inches long and $1\frac{1}{4}$ inches wide.
    What is its area?

4   The diagram is the plan of the floor of a workshop.
    The length of each side is correct to 1 decimal place.
    Find the area of the workshop floor.

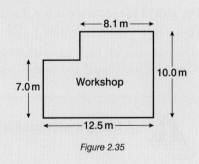

Figure 2.35

5   The sketch shows the plan
    of a car park.
    Find its area.

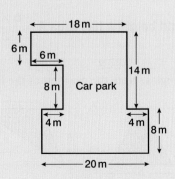

Figure 2.36

6 A garden consists of a rectangular lawn surrounded by a flowerbed as shown in the sketch.

Calculate the area of:
a the whole garden
b the lawn
c the flowerbed.

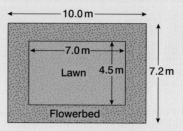

Figure 2.37

7 The sketch shows the design of a window frame.
a Calculate, in cm², the area of:
i the glass
ii the wooden surround.
b Convert each answer to m².
$\left(\begin{array}{l}\text{Remember: } 1\,m^2 = 100^2\,cm^2 \\ \qquad\qquad\quad = 10\,000\,cm^2\end{array}\right)$

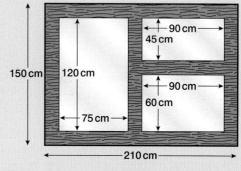

Figure 2.38

LEVEL 3

8 To give a particular output, solar panels are required to have an area of 2.4 m².
Find the required length of a rectangular panel if its width is:
a 0.6 m   **b** 0.8 m   **c** 1 m   **d** 1.2 m   **e** 1.5 m

A triangle's area depends on its base and height

The rectangle has been cut in half along one of its diagonals to produce two triangles.

The area of the rectangle is 6 × 4 = 24 cm².
Each triangle has area 12 cm² (= 24 ÷ 2)

In the formula for the area of a triangle, 'length' is replaced with 'base' and 'width' with 'height'.

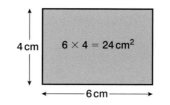
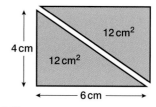

Figure 2.39

> ⚠️ **A triangle with base *b* and height *h* has area $A = \dfrac{b \times h}{2}$**

This diagram shows three different triangles. Note the position of the height *h* in each triangle.

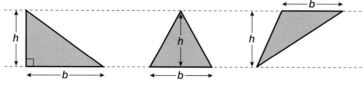

hint
The height must be at right-angles to the base

Figure 2.40

Example

The diagram shows the plan of a garden.
The plan consists of a rectangle
and a triangle.

Calculate the area of the garden.

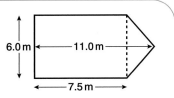

Figure 2.41

Here's how...

Treat the rectangle and triangle separately.

Area of rectangle $= 6 \times 7.5 = 45\ \text{m}^2$

Base of triangle b $= 6\ \text{m}.$

Height of triangle h $= 11 - 7.5 = 3.5\ \text{m}$

Area of triangle $= \dfrac{b \times h}{2} = \dfrac{6 \times 3.5}{2} = \dfrac{21}{2}$

$\qquad\qquad\qquad = 10.5\ \text{m}^2$

Total area of garden $=$ area of rectangle $+$ area of triangle

$\qquad\qquad\qquad\quad = 45\ \text{m}^2 + 10.5\ \text{m}^2$

$\qquad\qquad\qquad\quad = 55.5\ \text{m}^2$

Answer: Area $= 56\ \text{m}^2$ (to 2 s.f.)

hint

Quick method:
$6 \times 7.5 = 3 \times 15 = 45$
(or use the decimal
methods in ToTT.3)

hint

Quick method:
$6 \times 3.5 = 3 \times 7 = 21$

hint

Round to 2 s.f.
(same as least accurate
measurement given).
(See ToTT.8)

Now try these...

1 The sketch shows the dimensions
of a variety of flags.
Find the area of each flag.

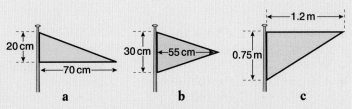

Figure 2.42

2 The diagram shows the wall at one end of
a house. The wall is to be waterproofed
at a cost of £5 per square metre.
Find the total cost.

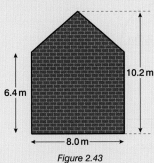

Figure 2.43

3 The diagram shows the sketch made by a
surveyor of a piece of land. Find the area of
the land, giving your answer in hectares.
(1 hectare = 10 000 square metres)

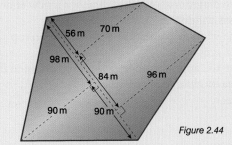

Figure 2.44

4 Kate makes an aluminium bookshelf
with two rectangular sides and two
triangular ends. Calculate the total
area of aluminum she uses.

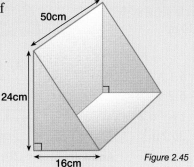

Figure 2.45

A circle is a plane shape

A **radius** r is any straight line from the centre to the edge of the circle.

A **diameter** d is a straight line passing through the centre joining opposite points on the circle.

The perimeter of a circle is called the **circumference**.

A **semicircle** is half a circle.

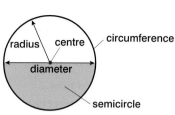

Figure 2.46

 A circle with radius r has: Circumference $C = 2\pi r$ (or $C = \pi d$)
and Area $A = \pi r^2$

$\pi = 3.141592654...$ An approximate value of 3 or 3.1 for π may be used in some calculations.

Example

A circular plate has radius $r = 10$ cm.
Use the approximation $\pi \approx 3.1$ to estimate:
a the circumference   **b** the area of the plate.

Here's how...

a $C = 2\pi r \approx 2 \times 3.1 \times 10$
$= 62$ cm

Answer: Circumference $= 62$ cm

b $A = \pi r^2 \approx 3.1 \times 10^2$
$= 3.1 \times 100$
$= 310$ cm^2

Answer: Area $= 310$ cm^2

hint
For decimal methods
see ToTT.3.

checkpoint
Check using $\pi \approx 3$:
$C \approx 2 \times 3 \times 10 = 60$ cm
and
$A \approx 3 \times 10^2 = 300$ cm^2

Example

The diagram shows the frame for a new stained glass window.
The window consists of a semicircle on top of a rectangle.
Use the approximation $\pi \approx 3.1$ to estimate the area of glass required.

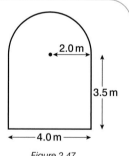

Figure 2.47

Here's how...

Treat the rectangle and semicircle separately.
Area of rectangle $= 4 \times 3.5 = 14$ m^2

Radius of semi-circle is 2 m.

Whole circle has area $\pi \times 2^2 \approx 3.1 \times 4 = 12.4$ m^2

Semicircle has area $12.4 \div 2 = 6.2$ m^2

Total area $= 14 + 6.2$
$= 20.2$ m^2

Answer: Area $= 20$ m^2 (to 2 s.f.)

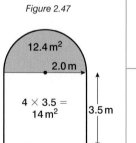

Figure 2.48

checkpoint
Check using approximations:
Rectangle $A \approx 4 \times 4 = 16$ m^2
Circle using $\pi \approx 3$,
$A \approx 3 \times 2^2$
$= 3 \times 4 = 12$ m^2

hint
Lengths and π were
given to 2 s.f.

Now try these...

Now try these...

1 The diameter of the London Eye is approximately 130 m. Estimate the length of its circumference using $\pi \approx 3.1$.

Figure 2.49

2 Measure the diameter of a 1p coin in millimetres. Using $\pi \approx 3.1$, estimate the coin's
   a circumference
   b area.

3 The radius of a bicycle wheel is 32 cm.
   a Estimate the circumference of the wheel using $\pi \approx 3.1$.
   b Approximately how many revolutions will the wheel make when the bicycle travels one kilometre?

4 The recipe for a cake is for a square cake tin with sides of length 20 cm. Matt only has circular cake tins with diameters 18 cm, 20 cm, 22 cm and 24 cm. Which of these tins has an area that is nearest in size to the square tin? (Use $\pi \approx 3.1$)

5 A 400 m running track is in the shape of a rectangle with two semicircular ends. The diameter of each end is 50 m. Calculate:
   a the length of each straight section of track
   b the area enclosed by the track.
   (Use π on your calculator then check your answers using $\pi \approx 3$.)

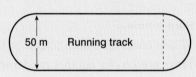

50 m   Running track

Figure 2.50

LEVEL 3

6 The diagram shows the dimensions of a paper clip when it is opened out.
   a Find, correct to 1 d.p., the total length of the wire used to make the paper clip.

Last year a manufacturer sold 50 thousand boxes, each containing 60 of these paper clips.
   b Find, in kilometres, the length of wire used by the manufacturer to produce these paper clips.

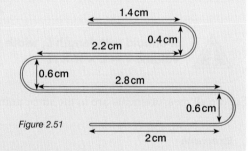

1.4 cm
0.4 cm
2.2 cm
0.6 cm
2.8 cm
0.6 cm
2 cm

Figure 2.51

7 A groundsman uses a trundle wheel to measure distances. Each turn of the trundle wheel measures 1 metre. Calculate the radius of the trundle wheel correct to the nearest centimetre.

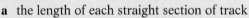

1 m

Figure 2.52

8 The cross section of the trunk of a tree is approximately circular with circumference of length 2.25 m. Assuming that the tree produces a tree ring approximately 0.5 cm wide each year, estimate the age of this tree.

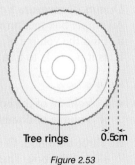

Tree rings   0.5 cm

Figure 2.53

9 **a** Make r the subject of the formula $A = \pi r^2$
   b A pipe is required to have a circular cross section with area 7 cm². Calculate the diameter of the pipe, correct to the nearest millimetre.

2.3.2 Volumes and solid shapes

A plane shape is flat. It has two dimensions – length and width. A solid is a three dimensional shape. It has length, width *and* depth (or height). Example of solids are:

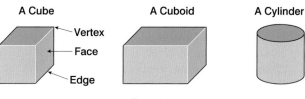

A Cube   A Cuboid   A Cylinder

Vertex
Face
Edge

Figure 2.54

Volume is measured in cubic units

The volume of a solid is the number of unit cubes that exactly fit inside the shape.

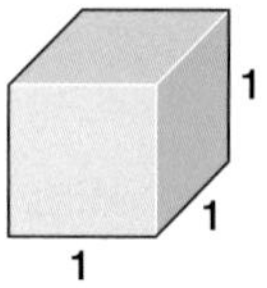

Figure 2.55

Example

A matchbox has length 4 cm, width 3 cm and height 3 cm.
Find the volume of the matchbox in cm^3.

Here's how...

12 unit cubes have been placed on the bottom of the box.
There is room for another two layers of unit cubes.

Volume = $3 \times 12 = 36$ unit cubes
        = $36\ cm^3$

Figure 2.56

Answer: $36\ cm^3$

checkpoint

Alternatively, use the formula:
$V = L \times W \times H$
$= 4 \times 3 \times 3 = 36\ cm^3$

 A cuboid with length *L*, width *W* and height *H* has volume $V = L \times W \times H$

Ensure all measurements are in the same units when working out a volume.

Example

Dominoes measure 5 cm by 2.4 cm by 5 mm.
An empty box measures 15 cm by 5 cm by 2.2 cm.
a Find the volume of: **i** the box   **ii** a domino.
b Is it possible to fit a set of 28 dominoes into the box?

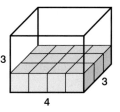

5 mm
2.4 cm
5 cm

Here's how...

a i Volume of the box: $15 \times 5 \times 2.2$ **Answer:** $165\ cm^3$
   ii Volume of domino: $5 \times 2.4 \times 0.5$ **Answer:** $6\ cm^3$

b Three dominoes fit in a line along the side of the box. There is room for 2 lines on the base and 4 layers. Number of dominoes = $3 \times 2 \times 4 = 24$

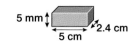

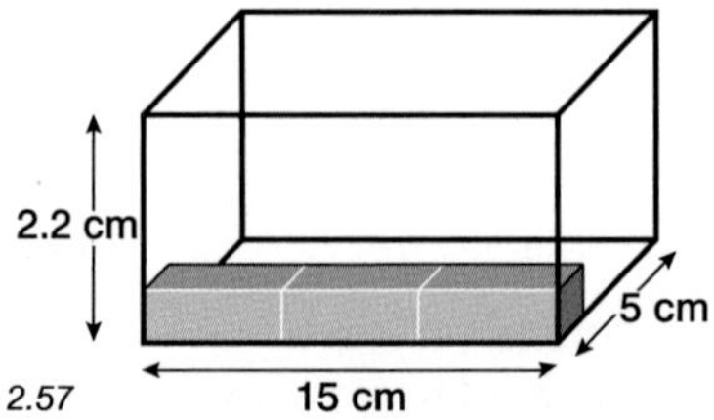

2.2 cm
5 cm
15 cm

Figure 2.57

Answer: 28 dominoes will not fit

hint

Convert 5 mm to cm:
$5 \div 10 = 0.5$ cm
(see Section 2.1.6)

checkpoint

Think about how they will fit in.
Take care!
Dividing volumes gives only a rough check:
$165 \div 6 = 28$ to the nearest whole number.

Now try these...

1 A packing case is 2 m long, 1.5 m wide and 1.2 m high. Calculate its volume.

2 A rectangular sandpit is 1.5 m long and 1.4 m wide. What volume of sand would be needed to fill the sandpit to a depth of 50 cm?

3 **a** A packet of ground coffee is 8 cm long, 5 cm wide and 15 cm high. Find its volume.
 b Greg uses 45 cm³ for each pot of coffee. How many pots of coffee can he make with a full packet?

4 **a** Use a unit centimetre cube to show that 1 cm³ = 1000 mm³
 b How many cm³ are there in 1 m³?
 c Convert: **i** 0.8 cm³ to mm³ **ii** 240 000 cm³ to m³.

5 Mandy's fish tank is 40 cm long, 30 cm wide and 25 cm deep.

 a Find its volume in litres.
 (1 litre = 1000 cm³)

 Mandy wants to buy some fish to put into the tank. She is advised to allow at least one and a half litres of water for each fish.

 b What is the maximum number of fish that she should buy?

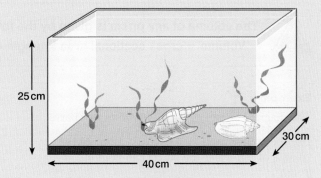

Figure 2.58

6 Boxes of drawing pins are 6 cm long, 5 cm wide and 2 cm deep. They are packed into a larger box that is 30 cm long, 24 cm wide and 10 cm deep. Find:
 a the volume of each box of drawing pins
 b the volume of the larger box
 c the maximum number of boxes of drawing pins that can be packed into the larger box.

7 A box contains sugar cubes that have edges of length 12 mm. The box is 12 cm long, 7.5 cm wide and 6 cm high. Find (using a calculator if you wish):
 a the volume of one sugar cube (in cm³)
 b the volume of the box (in cm³)
 c the maximum number of cubes that will fit into the box.

8 Helen has 120 audio tapes, each in a case that is 11 cm long, 7 cm wide and 16 mm deep.
 a Find the volume of each case (in cm³). Use a calculator if you wish.
 b Helen has three boxes. Their dimensions are:
  A 45 cm by 35 cm by 10 cm   **B** 45 cm by 30 cm by 12 cm   **C** 55 cm by 30 cm by 10 cm
  Which of these boxes would hold Helen's collection of tapes?

LEVEL 3

9 1000 litres of water are poured into a rectangular paddling pool that is 2.5 m long and 1.8 m wide. Find, to the nearest centimetre, the depth of the water. (1 litre = 1000 cm³.)

10 A carton of orange juice is designed to hold 1 litre when it is full. The carton is 10 cm long by 8 cm wide. How tall is it? (1 litre = 1000 cm³)

Prisms have a constant cross section

Look at the two wedges of cheese. A **cross section** has been made by slicing vertically through each wedge.
The cross section of the left hand wedge looks identical to its front face. The wedge has a **constant** cross section. It is called a **prism**. The right hand wedge is not a prism because it does not have a constant cross section.

Cubes, cuboids and cylinders are prisms. Everyday examples of these include a cereal box (cuboid), a baked bean tin (a cylinder or 'circular prism') and a Toblerone box (a triangular prism).

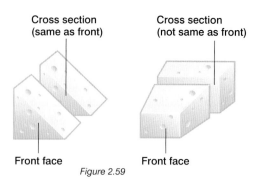

Cross section
(same as front)

Cross section
(not same as front)

Front face   Front face

Figure 2.59

 The volume of *any* prism is given by the formula:
Volume = cross sectional area × length or $V = A \times L$

Example

The diagram shows a cylindrical rainwater barrel.

Use the approximation $\pi \approx 3.1$ to estimate the volume of rainwater the barrel can hold in:
a cm^3
b litres (1 litre = 1000 cm^3)

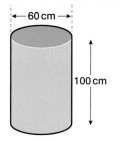

← 60 cm →

100 cm

Figure 2.60

Here's How...

a The cross section of the barrel is a circle of radius $r = 30$ cm.
Cross sectional area $A = \pi r^2 \approx 3.1 \times 30^2$
$\qquad\qquad = 3.1 \times 900$
$\qquad\qquad = 2790 \ cm^2$
Length (or height) $\qquad L = 100$ cm
Volume $\qquad\qquad V = A \times L \approx 2790 \times 100$
$\qquad\qquad\qquad = 279\,000 \ cm^3$
Answer: 280 000 cm³ (to 2 s.f.)

b Convert the answer to litres:
$279\,000 \ cm^3 = 279\,000 \div 1000$ litres
$\qquad\qquad = 279$ litres
Answer: 280 litres (to 2 s.f.)

hint
For decimal methods see ToTT.3. Check using
$\pi \approx 3, A \approx 3 \times 30^2$
$\qquad = 3 \times 900$
$\qquad = 2700 \ cm^2$

hint
The accuracy of the given measurements is not known. Round the answers to 2 s.f. because the value used for π is only correct to 2 s.f.

hint
Use all available figures in calculations. Round to 2 s.f. at the end.

Now try these...

1 A can of soup has a diameter of 7.8 cm and height 10 cm. Estimate the volume of the can using $\pi \approx 3$.

2 Figure 2.61 shows a wooden wedge used to prop open a door.
Find the volume of the wedge.

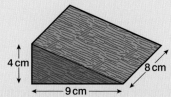

4 cm

8 cm

← 9 cm →

Figure 2.61

3 A cylindrical oil drum is 0.8 m high and has a diameter of 0.6 m. Estimate the volume of the can, in litres, using $\pi \approx 3$. (1 m^3 = 1000 litres).

4 A lean-to shed is constructed against the wall of a house.
Its dimensions are shown in figure 2.62.
 a Find the area of the cross section of the shed.
 b Calculate the volume of the shed.

5 A cylindrical coffee pot has a diameter of 12 cm and
height 20 cm.
 a Estimate the volume of the coffee pot using $\pi \approx 3$.

Matching coffee cups are also cylindrical with diameter 8 cm
and height 7 cm.
 b Approximately how many coffee cups can be filled from a
coffee pot full of coffee?

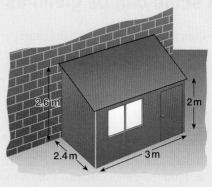

Figure 2.62

LEVEL 3

6 **a** The formula for the volume of a cylinder is $V = \pi r^2 h$ where r is the radius and h is the height.
Rearrange the formula to make the subject **i** h **ii** r
 b Cylindrical containers are designed to hold 100 litres of oil. (1 litre = 1000 cm³)
   i Find the height of a container whose radius is 20 cm.
   ii Find the radius of a container whose height is 20 cm.

7 The diagram shows the dimensions of a windowbox.
Calculate the number of 40 litre bags of compost
that are needed to fill the windowbox.

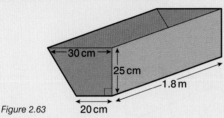

Figure 2.63

8 A packet of ready-made icing is in the
shape of a cuboid 12 cm long, 8 cm
wide and 2 cm deep. The icing is rolled
out to cover the top of a circular cake
whose diameter is 24 cm. Assuming
that all the icing is used, calculate the
thickness of the icing on the cake,
correct to the nearest millimetre.

2.3.3 Working with scales

When designing a house, an architect draws a plan. To make it fit on a piece of paper, he uses a **scale**.
The actual (or 'real life') dimensions of each room are scaled down. The result is a **scale drawing**.

Example

This is a scale drawing of a room.
The scale used is 1 centimetre = 2 metres.

By measuring each side, find the actual:
a length
b perimeter of the room.

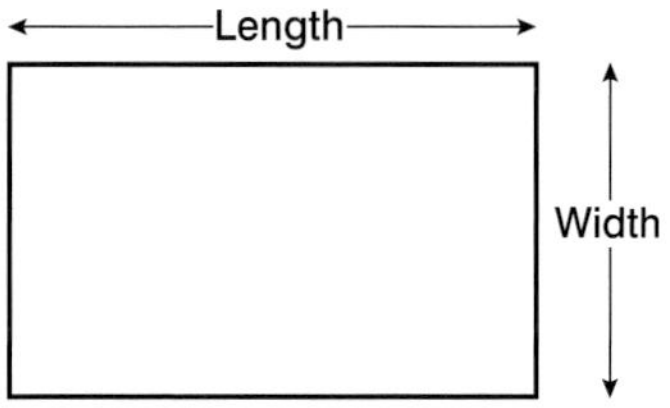

←——————Length——————→

Width

Figure 2.64

Here's how...

The scale 1 centimetre = 2 metres means each centimetre on the drawing
represents 2 metres in real life.
a Use a ruler to measure the length of the room. It is 3.6 cm.
   Actual length: $3.6 \times 2\,\text{m} = 7.2\,\text{m}$ **Answer:** 7.2 metres
b The width of the room measures 2.2 cm.
   Actual width: $2.2 \times 2\,\text{m} = 4.4\,\text{m}$
   Actual perimeter: $7.2 + 4.4 + 7.2 + 4.4 = 23.2\,\text{m}$ **Answer:** 23 metres (2 s.f.)

checkpoint

Check using
approximation:
$4 \times 2 = 8$

checkpoint

Check:
Perimeter of drawing
$= 2 \times 3.6 + 2 \times 2.2$
$= 7.2 + 4.4$
$= 11.6\,\text{cm}$
Actual perimeter
$= 2 \times 11.6$
$= 23 .2\,\text{m}$

A scale can be given as a ratio

If 1 cm represents 2 m then the ratio is:

1 cm : 2 m = 1 cm : 200 cm = 1 : 200

1 : 200 means '1 unit on the drawing = 200 units in real life'

200 is the *scale factor* (s.f.)

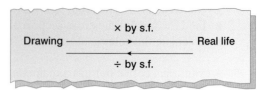

Drawing — × by s.f. → ÷ by s.f. ← Real life

Example

A map of Evesham is given below. The scale is 1 : 50 000.

a Find the direct distance from Twyford to Chadbury.

b Estimate the length of the river from Leicester Tower to Parks Farm.

c What lies on the A4184 approximately 3 km south of the station?

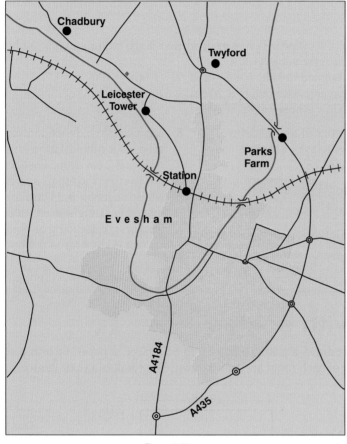

Figure 2.65

> **hint**
>
> Use a ruler to measure direct (straight line) distances. Use a piece of cotton or string to measure curved lengths.

Here's how...

a Direct distance on the map = 4.2 cm.
Actual distance: 4.2 × 50 000 = 210 000 cm = 2100 m   **Answer:** 2.1 km

b Length of this section of river on the map is approximately 12 cm.
Actual distance: 12 × 50 000 = 600 000 cm = 6000 m   **Answer:** 6 km

c Actual distance: 3 km = 3000 m = 300 000 cm
Distance on the map: 300 000 ÷ 50 000 = 6 cm

On the map 6 cm south of the station, there is a roundabout on the A4184

Answer: Roundabout

> **hint**
>
> × 50 000
> Map —————— Real life
> ÷ 50 000
> Give answers in 'sensible' units.

Now try these...

1 The scale diagram shows the plan of a swimming pool.
1 centimetre represents 5 metres.
  a By taking measurements from the diagram find:
    i the length of the pool,
    ii the width of the pool,
    iii the width of each training lane.
  b Calculate
    i the perimeter of the pool,
    ii the area of the pool,
    iii the area of each training lane.

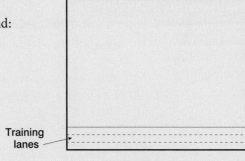

Scale: 1 cm represents 5 m

Training lanes

Figure 2.66

2 The scale drawing shows the plan of the rooms in a youth club.
  a By taking measurements from the diagram find the length and breadth of each room.
  b Calculate the area of the floor in each room.
  c Approximately what fraction of the floor area is the storeroom?

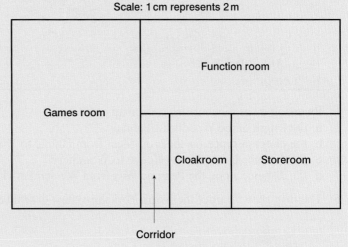

Scale: 1 cm represents 2 m

Function room

Games room

Cloakroom   Storeroom

Corridor

Figure 2.67

3 Measure the dimensions of the room in which you are working.
Construct a scale drawing of the room using a scale of your choice.

4 This plan of a kitchen is drawn to a scale of 1:50.
By taking measurements from the plan find (in metres):
  a the length and breadth of the kitchen,
  b the width of the window,
  c the width of the doors,
  d the width of the work surfaces,
  e the length and width of the table,
  f the length and width of the kitchen sink.

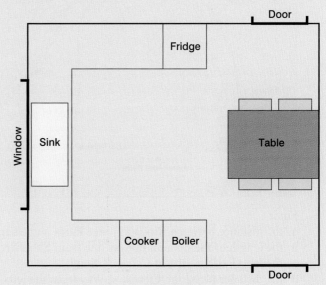

Door

Fridge

Window

Sink

Table

Cooker   Boiler

Door

Figure 2.68

5 The map shows part of the Peak District drawn to a scale of 1 : 100 000.

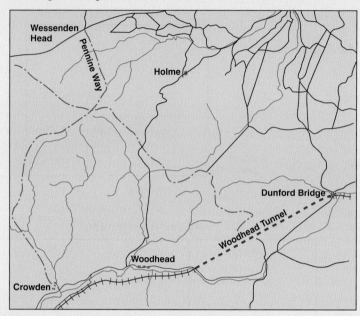

Figure 2.69

By taking measurements from the map estimate:
a the length of the Woodhead Tunnel
b the direct distance (as the crow flies) from Holme to Dunford Bridge
c the distance by road from Holme to Woodhead
d the distance along the Pennine Way from Wessenden Head to Crowden.

6 The map shows part of the North Yorkshire coast drawn to a scale of 1:25 000.

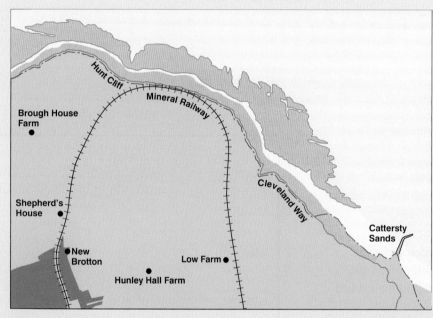

Figure 2.70

Find:
a the distance between Brough House Farm and Shepherd's House
b the length of the mineral railway line from Shepherd's House to Low Farm
c the length of the jetty at Cattersty Sands
d the distance along the Cleveland Way from the jetty at Cattersty Sands to Hunt Cliff
e the farm that lies approximately $1\frac{1}{8}$ kilometres east of New Brotton.

7 Write each of the following scales as a ratio in the form 1:*n*.
 a 1 cm represents 2 km   **b** 1 cm represents 25 km   **c** 1 cm represents 500 m
 d 2 cm represents 1 km   **e** 1 inch represents 1 mile   **f** 1 inch represents 4 miles

8 Measure the dimensions of the rooms in your house.
 Draw a scale drawing of each floor, using a scale of your choice.
 Give your scale in the form 1:*n*.

LEVEL 3 Enlargements increase all lengths by a scale factor

Each side of the *Enlarged* print is twice the length of the corresponding side in the *Regular* print.

The scale factor of the enlargement is 2.

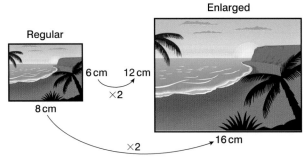

Figure 2.71

 Scale factor = $\dfrac{\text{length of side in new shape}}{\text{length of corresponding side in original shape}}$

Example

The diagram shows an illustration in a text book. The illustration is enlarged using a photocopier. Find:
a the scale factor of the enlargement,
b the width of the photocopy.

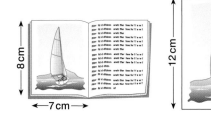

Figure 2.72

hint

Scale factor =
$\dfrac{\text{length of side in new shape}}{\text{length of corresponding side in original shape}}$

Here's how...

a Compare corresponding sides:
 Scale factor = $\dfrac{12 \text{ cm (photocopy)}}{8 \text{ cm (text book)}}$

 Answer: Scale factor = 1.5

b Width of photocopy = $7 \times 1.5 = 10.5$ cm

 Answer: width = 10.5 cm

hint

In an enlargement each side is multiplied by the same scale factor.

The area scale factor is the square of the length scale factor

A rectangle with length 4 cm and width 2 cm
is enlarged by scale factor 3.

Area of smaller rectangle = 4 cm × 2 cm = 8 cm².

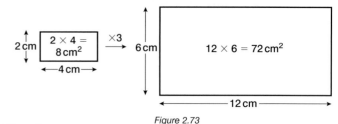

In the enlargement each side is multiplied by 3.
Area of enlarged rectangle is $3 \times 3 = 9$ times
greater than the area of the smaller rectangle.

Figure 2.73

As a numerical check: $\dfrac{\text{area of enlargement}}{\text{area of smaller rectangle}} = \dfrac{72 \text{ cm}^2}{8 \text{ cm}^2} = 9 \ (= 3^2)$

 If lengths are enlarged by a scale factor *s* then areas are
enlarged by a scale factor of s^2

Example

A design for a company logo is a circle touching the edges
of a square. The square has side 20 cm.
a Use the approximation $\pi \approx 3.1$ to estimate the area of
the circle.

The design is enlarged by scale factor 5.
b Estimate the area of the enlarged circle in cm².

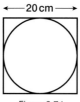

Figure 2.74

hint

Diameter of circle = 20 cm
Radius of circle = 10 cm

Here's how...

a $r = d \div 2 = 10$ cm

Area $A = \pi r^2 \approx 3.1 \times 10^2 = 3.1 \times 100$
$= 310 \text{ cm}^2$

hint

Take care to use *r*
not *d* in the formula.

checkpoint

Makes sense: answer is
less than the area of the
square (20 × 20 = 400).

Answer: Area $\approx 310 \text{ cm}^2$

b Scale factor $s = 5$. Area scale factor $= s^2 = 5^2$
$= 25$

Area of enlarged circle $\approx 25 \times 310 \text{ cm}^2$
$= 7750 \text{ cm}^2$

Answer: Area $\approx 7800 \text{ cm}^2$ (2 s.f.)

hint

π is correct to only 2 s.f.
Round answers to 2 s.f.

The volume scale factor is the cube of the length scale factor

The first cuboid has length 2 cm, width 1 cm and height 3 cm.
It is enlarged by scale factor 2.

The volume of the smaller cuboid is

$2 \text{ cm} \times 1 \text{ cm} \times 3 \text{ cm} = 6 \text{ cm}^3$

In the enlarged cuboid, each side is multiplied by 2.
The *volume* of the enlarged cuboid is $2 \times 2 \times 2 = 8$ times
greater than that of the smaller cuboid.

Figure 2.75

Numerical check: $\dfrac{\text{volume of the enlargement}}{\text{volume of smaller cuboid}} = \dfrac{4 \times 2 \times 6}{2 \times 1 \times 3} = \dfrac{48}{6} = 8 \ (= 2^3)$

If lengths are enlarged by a scale factor s then volumes are enlarged by a scale factor of s^3

Example

The diagram shows the design for a box.

a Calculate the volume of the box.

A change in the specifications require each length to be increased by 10%.

b Find the scale factor for the enlargement.

c Calculate the volume of the enlarged box.

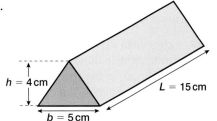

$h = 4\,cm$ $L = 15\,cm$ $b = 5\,cm$

Figure 2.76

Here's how...

a The box is a triangular prism.

Cross sectional area $A = \dfrac{b \times h}{2} = \dfrac{5 \times 4}{2} = \dfrac{20}{2}$
$$= 10\ cm^2$$

Volume $V = A \times L = 10 \times 15 = 150\ cm^3$

Answer: Volume = 150 cm³

> **hint**
> Volume of prism = Cross-sectional area × length

b A 10% increase means each new length is

$110\% \times$ original length $= 1.1 \times$ original length

Answer: Scale factor = 1.1

> **hint**
> For percentages see ToTT.5 and Section 2.1.3.

c Scale factor for length is s $= 1.1$
Scale factor for volume is $s^3 = 1.1^3 = 1.331$
Enlarged volume $= 1.331 \times 150\ cm^3$
                       $= 199.65\ cm^3$

Answer: Enlarged volume = 200 cm³ (nearest cm³)

> **hint**
> Use a calculator here – see ToTT.7

> **hint**
> Round the answer sensibly.

Now try these...

1 A print of a photograph is 4 inches wide and 6 inches high. In an enlargement of the print the width is 5 inches.
Find
  a the scale factor of the enlargement
  b the height of the enlargement.

← 4 in →   ← 5 in →

6 in

Figure 2.77

2 A photograph negative measures 36 mm by 24 mm.
When a print is made of the negative its longer side measures 18 cm.
  a Find the scale factor of the enlargement.
  b Calculate the length of the shorter side of the print.

3 A diagram is enlarged with scale factor 4.
  a What is the area scale factor?
  b If the original diagram had an area of 7.5 cm², find the area of the enlargement.

4 A photocopier is set to increase the length and width of a drawing to 120% of its original size.
  a What is **i** the length scale factor   **ii** the area scale factor?
  b The length of the original drawing was 5 cm. What is the length of the enlargement?
  c The area of the original drawing was 20 cm². What is the area of the enlargement?

5 A scale model of the wing of an aircraft is used in tests. The model is made using a scale of 1:10.
  a What is the area scale factor? Give your answer in the form 1:n.
  b If the area of the model wing is 0.5 m², what is the area of the actual wing?

6 A confectioner makes bars of chocolate that are 10 cm long, 6 cm wide and 2 cm thick.
  Special bars are made at Christmas whose dimensions are 50% bigger than the usual bars.
  a Find the length, width and thickness of the special bars.
  b Calculate the volume of **i** the usual bars   **ii** the special bars.
  c What is the volume scale factor?

7 An ice cream van has a model of an ice cream cone fixed to its side.
  It is an enlargement, five times the height of an actual cone.
  How many times larger is the volume of the model than the volume of an actual ice cream cone?

8 A large box of breakfast cereal is an enlargement of a standard box. The length scale factor is 1.2.
  a Find:
    i the area scale factor
    ii the volume scale factor.
  b The front of the standard box has an area of 450 cm². Calculate the area of the front of the large box.
  c The volume of the standard box is 2700 cm³. Calculate the volume of the large box.

9 A tin of cocoa has a height of 10 cm and contains 450 cm³ of cocoa when full.
  The manufacturer makes a tin that is an enlargement of the smaller tin with scale factor 1.5.
  What is:
  a the height
  b the volume
  of the larger tin?

10 A shop sells two different sized bottles of olive oil. The smaller bottle contains 400 ml and the larger bottle contains a litre.
  a What is the volume scale factor?
  b Calculate the length scale factor.
  c If the height of the smaller bottle is 10 cm, what is the height of the larger bottle?

2.3.4 Working with right-angled triangles

LEVEL 3   **A right-angled triangle has an angle of 90°**

The corners of a triangle are usually labelled with capital letters.
Angle A is the angle inside the triangle at corner A.

The longest side is called the **hypotenuse**.
It is always opposite the right angle.

If one of the angles (other than 90°) is given then the side opposite the given angle is called the **opposite** side.
The side against the given angle is called the **adjacent** side.

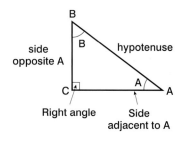

Figure 2.78

Example

In the right-angled triangle the given angle is 46°.
Label the sides of the triangle.

Here's how...

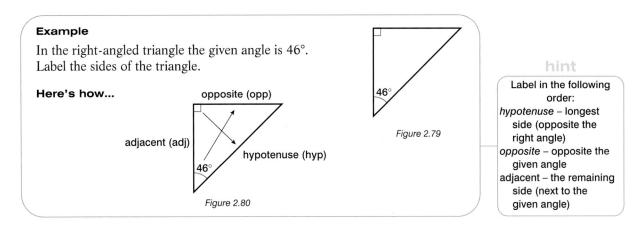

Figure 2.79

Figure 2.80

Now try these...

Label the sides of the triangles

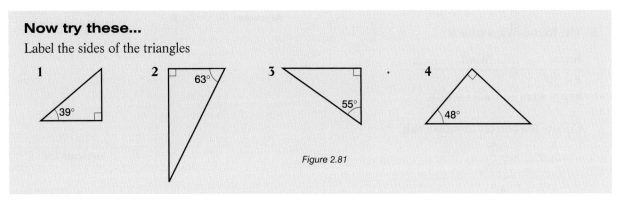

Figure 2.81

LEVEL 3   Using angles to find sides

If one side and one angle (other than 90°) are given, then *all* sides of the triangle can be found. In the diagram, the given angle is labelled as θ. Three formulae connect this angle to the sides of the triangle.

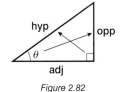

Figure 2.82

$$\sin \theta = \frac{\textbf{opposite}}{\textbf{hypotenuse}} \qquad \cos \theta = \frac{\textbf{adjacent}}{\textbf{hypotenuse}} \qquad \tan \theta = \frac{\textbf{opposite}}{\textbf{adjacent}}$$

The three formulae can be abbreviated to **soh¦cah¦toa** (take the first letter in bold in each formula).

For example, the first three letters '**soh**' mean $\sin \theta = \dfrac{\textbf{o}\text{pposite}}{\textbf{h}\text{ypotenuse}}$

The sin, cos and tan keys can be found on any scientific calculator.

Example

The triangle has hypotenuse of length 4 cm and given angle 30°

Calculate the length of:
a   side *BC*   **b**   side *AC*.

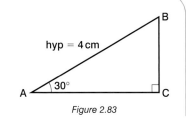

Figure 2.83

Here's how...

a Label the sides:
   Draw up a 'Know–Want' table:

hint

Method to find sides:
1 Label the sides.
2 Draw a 'Know-Want' table.
3 Choose the correct formula using **soh–cah–toa**.
4 Rearrange it to find the side you require.

Know	Want
$\theta = 30°$	
hyp = 4 cm	**o**pp ($= BC$)

Choose the correct formula: '**soh**'

$$\sin \theta = \frac{\mathbf{opp}}{\mathbf{hyp}}$$

$$\sin 30° = \frac{BC}{4} \qquad \text{giving } BC = 4 \times \sin 30° = 4 \times 0.5$$

Answer: $BC = 2$ cm

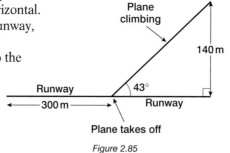

Figure 2.84

hint

Use the letters in the 'Know–Want' table as a guide: **soh–cah–toa**. Using angles to find sides is called trigonometry.

b The Know–Want table is:

Know	Want
$\theta = 30°$	
hyp = 4 cm	**a**dj ($= AC$)

Choose the correct formula: '**cah**'

$$\cos \theta = \frac{\mathbf{adj}}{\mathbf{hyp}}$$

$$\cos 30° = \frac{AC}{4} \qquad \text{giving } AC = 4 \times \cos 30°$$

$$= 4 \times 0.8660... = 3.464...$$

Answer: $AC = 3.5$ cm (to 1 d.p.)

hint

soh–**cah**–toa

checkpoint

In this example, the triangle in the question was drawn accurately. Check each answer by measuring with a ruler.

If the hypotenuse is neither given nor needed then you must use 'tan'.

Example

An aircraft starts its take off at one end of a runway. It accelerates 300 m along the runway before climbing at an angle of 43° to the horizontal. When directly above the end of the runway, the aircraft has altitude 140 m. Find the total length of the runway to the nearest metre.

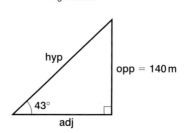

Figure 2.85

hint

Sketch a diagram showing all the information.

Here's how...

Label the sides of the triangle:

Know	Want
$\theta = 43°$	
opp = 140 m	**a**dj

Use tan: $\tan \theta = \dfrac{\mathbf{opp}}{\mathbf{adj}}$

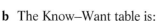

Figure 2.86

hint

soh–cah–**toa**

$$\tan 43° = \frac{140}{\text{adj}}$$

$$\tan 43° = \frac{140}{\text{adj}} \qquad \text{giving} \qquad \text{adj} = \frac{140}{\tan 43°} = \frac{140}{0.9325...} = 150.1...$$

Total length of runway = 300 + 150.1 = 450.1 m

Answer: Runway = 450 metres (nearest m)

Now try these...

1 An 8 metre ladder rests against the side of a house. The angle between the ladder and the ground is 75°. How high up the wall does the ladder reach?

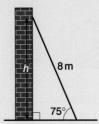

Figure 2.87

2 Sally starts from point P on the bank of a river. Directly opposite her on the other bank is a tree T. Sally walks along the river bank a distance 25 metres to point Q and measures the angle between QP and QT. It is 38°. Find the width of the river, correct to the nearest metre.

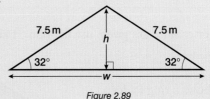

Figure 2.88

3 The diagram shows a roof support with sides of length 7.5 metres at an angle of 32° to the horizontal.
Calculate
 a the height *h*
 b the width *w* of the roof support.

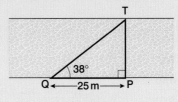

Figure 2.89

4 A ramp is required to allow wheelchair access to a building. The ramp is designed to rise at an angle of 15° to the horizontal. Its height is 20 cm.
Calculate $\quad$ **a** the width of the ramp, *w* $\qquad$ **b** the length of the ramp, *l*.

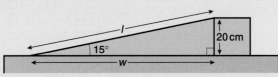

Figure 2.90

5 A children's slide is 8 m long and inclined at 42° to the horizontal.
 a Find the height of the slide, *h*, in metres correct to 2 decimal places.

The ladder to the top of the slide is inclined at 70° to the horizontal.
 b Calculate the length of the ladder, *l*, in metres correct to 1 decimal place.

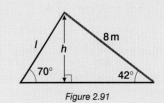

Figure 2.91

LEVEL 3 **Using sides to find angles**

If two sides of a right-angled triangle are given then all its angles can be found.
Angles are found using one of the keys $\boxed{\sin^{-1}}$, $\boxed{\cos^{-1}}$ and $\boxed{\tan^{-1}}$ on your
calculator. The $\boxed{\sin^{-1}}$ key has the reverse effect of the [sin] key.

For example, $\sin 30° = 0.5$ and $\sin^{-1} 0.5 = 30°$.

checkpoint

Use your calculator to
check $\sin^{-1} 0.5 = 30°$.
(The degree sign does not
appear on the screen.)

Example

A ramp is fitted to a ferry to
allow cars to disembark. The
ramp is 1 metre in height and
has slant length 3 metres.
Find the angle made by the
ramp with the ground to the
nearest whole degree.

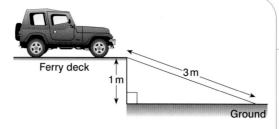

Figure 2.92

hint

Method to find angles:
1 Label the sides.
2 Draw a 'Know–Want'
  table.
3 Choose the correct
  formula using
  soh–cah–toa.
4 Divide the sides to get a
  decimal.
5 Use the [$\sin^{-1}$], [$\cos^{-1}$]
  or [$\tan^{-1}$] key to find the
  angle.

Here's how...

Label the sides:

Know	Want
opp = 1 m	
hyp = 3 m	θ

$\sin \theta = \dfrac{\text{opp}}{\text{hyp}}$ gives: $\sin \theta = \dfrac{1}{3} = 0.333...$

$\theta = \sin^{-1} 0.333... \quad = \quad 19.47...°$

Figure 2.93 (opp = 1m, hyp = 3 m, θ)

hint

soh–cah–toa

hint

Press the [$\sin^{-1}$] key.

Answer: Angle = 19° (nearest degree)

If only the two shorter sides are given, use the [$\tan^{-1}$] key to find an angle.

Example

The diagram shows a ring folder
laid on one side.

Find   **a** angle A
       b angle B

Give each answer to the nearest degree.

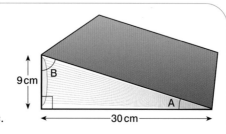

Figure 2.94

Here's how:

a Label the sides:

Know	Want
opp = 9 cm	
adj = 30 cm	θ

$\tan \theta = \dfrac{\text{opp}}{\text{adj}}$ gives: $\tan \theta = \dfrac{9}{30} = 0.3$

$\theta = \tan^{-1} 0.3 \quad = \quad 16.699...°$

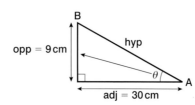

Figure 2.95 (opp = 9cm, hyp, adj = 30 cm, θ)

hint

soh–cah–**toa**

hint

Press the [$\tan^{-1}$] key.

checkpoint

Makes sense: 17° looks
reasonable.

Answer: A = 17° (nearest degree)

b Label the sides:

Know	Want
opp = 30 cm **a**dj = 9 cm	θ

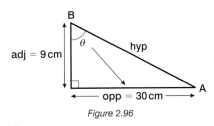

Figure 2.96

$$\tan \theta = \frac{\text{opp}}{\text{adj}} \qquad \text{gives:} \qquad \tan \theta = \frac{30}{9} = 3.333\ldots$$

$$\theta = \tan^{-1} 3.333\ldots = 73.30°$$

Answer: B = 73° (nearest degree)

Now try these...

1 A road sign says that a hill has a gradient of 1:5. This means that the road rises 1 m for every 5 m it goes horizontally. Calculate the angle the road makes with the horizontal.

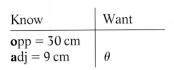

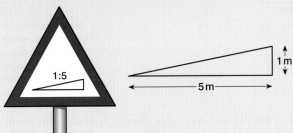

Figure 2.97

2 A skier descends by 50 m when she skis down a slope of length 180 m. Calculate the angle, θ, made by the slope with the horizontal.

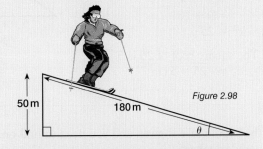

Figure 2.98

3 An aircraft is flying at an altitude of 3000 m. When it is 5 km from the start of a runway it starts its descent for landing.
At what angle to the horizontal should the aircraft descend to touch down at the beginning of the runway?

Figure 2.99

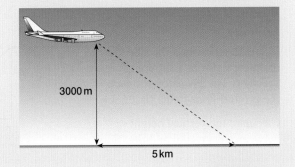

4 The diagram shows a conservatory attached to the wall of a house. The height of the front of the conservatory is 2 m and the height of the back is 2.5 m. The sloping roof is 3 m long.

Find θ, the angle between the roof and the vertical.

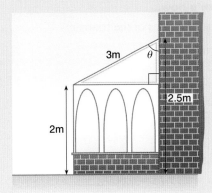

Figure 2.100

LEVEL 3 In a right-angled triangle the *longest* side squared equals the *sum* of the two shorter sides squared

Pythagoras' theorem gives a formula that connects the three sides of a right-angled triangle.

The formula does not involve angles.

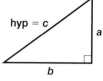

Figure 2.101

 Formula: $c^2 = a^2 + b^2$ where c = hypotenuse

Example

The lengths of the two shortest sides of a right-angled triangle are 3 cm and 4 cm. Find the length of the hypotenuse.

Here's how...

Formula:
$$c^2 = a^2 + b^2$$
$$= 3^2 + 4^2$$
$$= 9 + 16$$
$$c^2 = 25$$

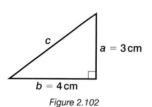

Figure 2.102

The square root of 25 is 5
$$c = \sqrt{25} = 5 \text{ cm}$$

Answer: $c = 5$ cm

If the hypotenuse is given and you need to find one of the two **shorter** sides, then you must **subtract** the squared numbers.

Example

Figure 2.103 shows the side view of an escalator in a shopping precinct. The escalator connects the ground floor to the first floor.

Calculate the vertical height between the ground and first floor.

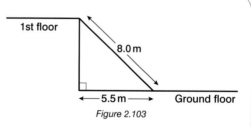

Figure 2.103

Here's how...

Label the sides of the triangle:

Formula:
$$a^2 + b^2 = c^2$$
$$a^2 + 5.5^2 = 8^2$$
$$a^2 + 30.25 = 64$$
$$a^2 = 64 - 30.25$$
$$a^2 = 33.75$$

Figure 2.104

Take the square root of 33.75 to find a.
$$a = \sqrt{33.75} = 5.809\ldots$$

Answer: Height = 5.8 metres (2 s.f.)

Now try these...

1 A mast is supported by two wires attached to the top and to points on the ground 20 m from its base (as shown in the diagram). The height of the mast is 30 m.
Calculate the length of each cable.

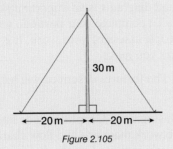

Figure 2.105

2 A gate is strengthened by fixing a strut along each of its diagonals. The gate is 2 m long and 1.2 m high.
Find the length of each strut.

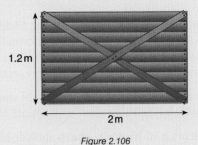

Figure 2.106

3 Ken takes a shortcut along the path from the gate to the bus stop. Calculate how much further it would have been if he had walked around the edges of the field.

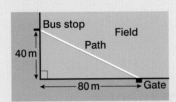

Figure 2.107

4 A tent is 2.8 m wide and its sloping sides are 2.4 m long. Calculate the height of the tent.

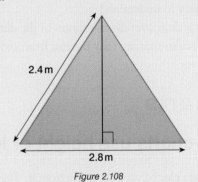

Figure 2.108

5 The sketch shows part of a rollercoaster ride between two points, P and Q at the same horizontal level. The highest point in this section is 20 m above PQ, the length of the ramp is 55 m and the length of the drop is 30 m.
Find the horizontal distance between P and Q.

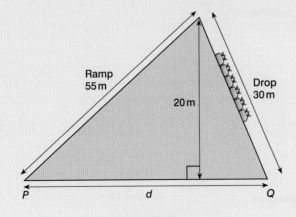

Figure 2.109

6 The cross section of a road tunnel is part of a circle of radius 4 m. The width of the tunnel at road level is 6 m.
Calculate its height, h, correct to 1 decimal place.

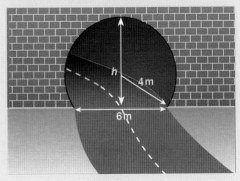

Figure 2.110

Charts, Graphs and Diagrams

Obtaining information and presenting findings using charts, graphs and diagrams is an important skill. Exam questions may require you to read information from a graph or chart. To complete your portfolio you must present your findings in a way that can be easily understood by the people assessing your work. It is vital that you follow the conventions for data presentation and justify the choices you make.

This section describes the main types of statistical charts and graphs. The features described will help you decide which type is the most appropriate for your data.

In Application of Number, **charts** include bar-charts, pie-charts and histograms. **Graphs** include line graphs and scatter graphs. **Diagrams** include scale drawings, maps, plans, network diagrams and flow charts. Any scale used for a drawing, map or plan must be sensibly chosen and clearly indicated on the diagram.

Every chart, graph and diagram that you draw must:

- include a title and labels
- be easy to understand
- give a clear overall impression of the data
- use scales that are easy to read from (whenever scales are used)

Charts

Bar-charts use height or length to show frequency

The bars can be vertical or horizontal. The length of the bar represents frequency (i.e. the number of items in the group).

Ideal for:

- displaying qualitative or discrete data
- showing individual parts of a total

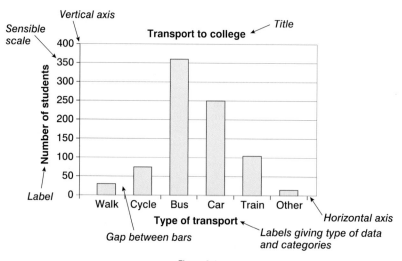

Figure 3.1

Special types of bar-charts include:

Comparative (or Compound) Bar-charts

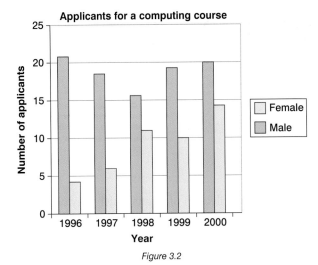

Figure 3.2

Component Bar-charts

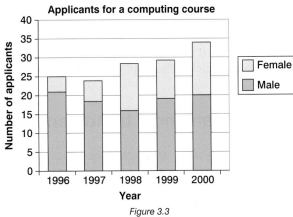

Figure 3.3

Ideal when:

- comparison between categories is more important than total values

Ideal for showing:

- the total number in each group
- how the total is made up from different categories

Histograms use area to show frequency

A histogram is used for continuous data or grouped discrete data. The bars are joined. Provided they are of equal width, the length of each bar represents the number of items in a given interval.

The histograms below summarise information about fish caught by a fishing boat over a 30 day period.

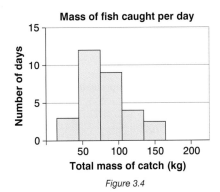

Figure 3.4

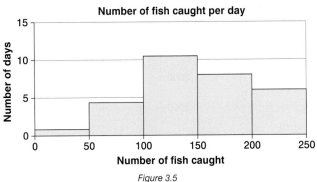

Figure 3.5

Pie-charts use angles to show proportion

A pie-chart is a circle divided into sectors representing different categories of data.

Ideal for showing:

- qualitative or quantitative data
- the size of each group as a proportion of the whole
- at a glance which is the largest or smallest category

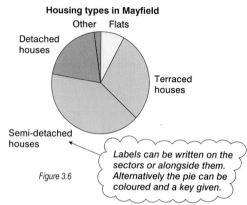

Figure 3.6

Labels can be written on the sectors or alongside them. Alternatively the pie can be coloured and a key given.

Graphs can show trends and relationships

A graph consists of a pair of axes (one horizontal, one vertical) and a collection of points which may or may not be joined. Each axis must use a suitable scale and be clearly labelled with appropriate units.

Line graphs

Each point is joined to the next using a straight line.

Ideal for:

- showing how continuous data varies, usually over a period of time
- comparing two or more sets of data
- identifying trends and making predictions

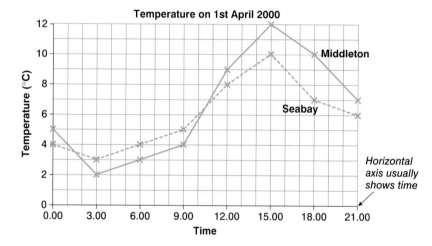

Horizontal axis usually shows time

Figure 3.7

Scatter graphs

Scatter graphs help to show when two quantities might be related. Each scale is marked with a quantity and points plotted. If the points lie within a narrow band there is strong 'correlation' between the quantities.

Ideal for:

- illustrating a relationship between two variables – in this case, *negative* correlation – the older the car, the less its value
- finding a formula relating the variables using the line of best fit

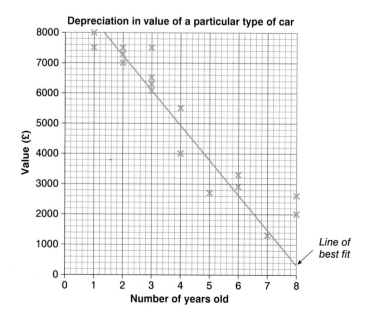

Line of best fit

Figure 3.8

Organising Data

3.2.1 Classifying Data

Data is useful information collected with a purpose in mind. For example, suppose we want to investigate the use of mobile phones. The table shows the type of information we might collect.

Information	Examples of data collected		
various makes of phones	*Lemon*	*Fone-A-Friend*	*11-Squared*
peak time rate (pence per minute)	40	35	30

Table 3.1

There are two main types of data:

1 Quantitative: data consisting of numbers (e.g. peak time rates)

2 Qualitative: non-numerical data (e.g. types of mobile phones)

It is important to distinguish between numerical and non-numerical data as this affects the way in which results can be presented.

Quantitative data consists of numbers

Quantitative data can itself be split into two categories: **discrete** and **continuous**

i Discrete data

Jack has in his pocket a standard UK coin with value between 5p and 50p.
The coin can only take a limited number of values.

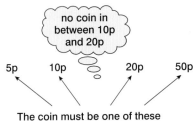

Figure 3.9

There are no other UK coins between 5p and 50p
Money is **discrete** data.

ii Continuous data

Suppose a tree has height between 10 metres and 11 metres.

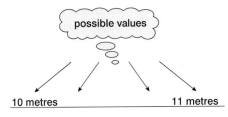

Figure 3.10

The tree could have a height of 10 metres, 11 metres or **any value between** these two measurements. Height is **continuous** data.

- **Discrete data:** can take only a limited number of values in a given range e.g. shoe sizes 8, $8\frac{1}{2}$, 9, $9\frac{1}{2}$ etc.
- **Continuous data:** can take an *unlimited* number of values within a given range, usually comes from taking measurements (e.g. foot length)

Qualitative data is non-numerical data

Non-numerical data is often descriptive.
In a road traffic survey, the first four makes of car to pass a check point were

Ford, Rover, Peugeot, Vauxhall

The make '*Ford* ' is a description, not a number. This data is **qualitative**.

Now try these...

1 Data is collected at a petrol station. Classify each type as discrete, continuous or qualitative.
  a The quantity of petrol taken from a pump
  b The price (in pence) of petrol per litre
  c The number of cars arriving at the petrol station
  d The different types of petrol selected by customers
  e The different number plates of each car arriving at the garage
  f The time taken for cars to go through the car wash

2 Data is collected in a library. Classify each type of data as discrete, continuous or qualitative.
  a The number of books on a shelf
  b The length of a shelf
  c The library codes for each book
  d The amount of money collected in fines in a day
  e The number of days a book is overdue

3 Data is collected in a clothes store. Classify each type of data as discrete, continuous or qualitative.
  a The shirt size of a customer
  b The waist measurement of a customer
  c The make of each shirt sold
  d The length of a sales assistant's measuring tape

3.2.2 Frequency tables

When data values are listed one by one, the data is said to be in **raw** form. Raw data is disorganised and difficult to understand. Trends and patterns are easier to see once the data has been processed in some way.

Frequency tables are a way of organising data into a more manageable form. The **frequency** of a value is the number of times the value appears. It is found most easily by counting with **tallies**.

Example

A supermarket stocks four brands of toothpaste:
Ultra-Fresh (U), White Bright (W), Condition Mint (C), McCavaties (M).
The table shows the sales of the brands of toothpaste in a one hour period.

U	M	C	W	U	W	C	W	C
M	W	C	U	W	U	M	C	W

Summarise the data using a frequency table. Identify the leading brand.

Here's how...

It is difficult to tell *at a glance* which is the most popular brand.

1 Draw up a table with four rows, one for each brand.
2 Work through the raw data *one item at a time*, placing a tally mark '|' in the appropriate row. Strike out that item from the list.
3 Record every fifth tally as ⊥⊦⊦⊤ to make counting easier. The table below shows the complete frequency table.

Brand	Tally	Frequency f
Ultra-Fresh (U)	||||	4
White Bright (W)	⊦⊦⊦⊤ |	6
Condition Mint (C)	⊦⊦⊦⊤	5
McCavaties (M)	|||	3
		Σf = 18

Table 3.2

Answer: The most popular brand is 'White Bright'

> **hint**
> The letter f stands for frequency.

> **checkpoint**
> Σf stands for the total of all the frequencies.
> Σf = 18 because 18 tubes of toothpaste were sold.

Frequency tables can also be used to organise quantitative data. For example, the number of mobile phone calls made by 20 students during one day is recorded below:

Raw Data

| 9 | 9 | 8 | 5 | 5 | 7 | 8 | 5 | 8 | 7 |
| 5 | 8 | 9 | 5 | 9 | 7 | 9 | 5 | 7 | 8 |

The data has been organised into a frequency table with values ranging from five to nine inclusive:

Again, the total frequency Σf must equal the total number of students (20). Also, the frequency of a value which does not appear in the list (such as '6') must be recorded as zero and not left blank.

No. calls made	Tally	Frequency f
5	ⱧⱧⱦ Ⅰ	6
6		0
7	ⅠⅠⅠⅠ	4
8	ⱧⱧⱦ	5
9	ⱧⱧⱦ	5
		$\Sigma f = 20$

Table 3.3

Now try these...

1   A shop stocks five types of mobile phone: *Backch@t, Call Britannia, FreeCall, N-Guage, TalkAbout*
    The sales data are:

| B | C | F | N | C | T | B | T | C | N |
| N | C | F | T | B | F | C | B | N | C |

   a Organise the data using a frequency table.   b Identify the most popular make of phone.

2   The number of injections given by a nurse each day are:

| 2 | 4 | 7 | 6 | 1 | 3 | 4 | 3 | 6 | 1 |
| 7 | 2 | 4 | 7 | 3 | 6 | 3 | 7 | 3 | 7 |

   a Organise the data using a frequency table.
   b What is the least number of injections given in a day by the nurse?

3   The football team *Oldcastle Utd* play 36 matches in a season. 18 matches are played at their home ground, the remainder are played away from home. The number of goals they score in each match during the season are shown below:

1H	2A	0H	2H	0A	1A	2H	1A	0A	0H	1H	2A
0H	3A	4H	2H	1A	3H	1A	2H	1A	0A	3H	0A
2A	1H	2A	1H	4H	0A	1A	0A	2H	0A	1H	1H

Key: 1H = 1 goal scored at <u>Home</u>   2A = 2 goals scored <u>Away</u>

   a Draw up frequency tables for the: i number of home goals   ii number of away goals
   b The manager claims '*Oldcastle Utd* find it harder to score goals when playing away from home'. Is the manager correct? Give a reason for your answer.

3.2.3 Grouping data

Data can be grouped in intervals

When data ranges over a small number of values it makes sense to use one row for each value in the frequency table (see the *Mobile Phone* example above).

For data spread over a *large* range, one row per value would require a large number of rows, each having a low frequency. To avoid this problem, data should be ***grouped into intervals***. The interval '1–5', for example, can be used to include all the numbers from 1 to 5.

Example

28 students monitor how many calls they make on their mobile phones during one weekend.

12	5	9	3	11	4	12
13	16	7	14	12	9	10
1	15	14	8	8	20	14
15	10	12	6	17	3	11

Organise the data into a *grouped* frequency table using four intervals. Comment on the table.

hint
A frequency table *without* intervals would require 20 rows (numbered 1 to 20). No overall pattern would emerge from such a table

Here's how...

To construct a grouped frequency table:

- Select appropriate intervals:
  '1–5', '6–10', '11–15', '16–20' gives the required four intervals
- Tally each value, identifying the interval to which it belongs. The first value is 12. This lies between 11 and 15 and so belongs to the interval '11–15'. The second value is 5. This value belongs to interval '1–5' (intervals include their end values).

hint
The smallest value is 1 and the largest is 20. Split the range 1 to 20 into 4 equal intervals.

Here is the completed frequency table:

No. calls made	Tally	Frequency f
1–5	卌	5
6–10	卌 |||	8
11–15	卌 卌 ||	12
16–20	|||	3
		$\Sigma f = 28$

Table 3.5

hint
$\Sigma f = 28$ = number of students

Answer: Phones were used most frequently between 11 and 15 times

Intervals are useful because they allow the general picture to be seen. Their drawback is that only general statements can be made. From the grouped frequency table above, we cannot tell which is the most common number of calls made.

If you have to choose your own intervals, make each interval sufficiently wide to allow an overall pattern to be seen. This normally requires using between four and eight intervals if the amount of data is relatively small.

Remember: the narrower you make each interval, the more intervals you will need to use and the more data you will need in order to see a pattern.

Now try these...

1 On a particular Saturday, the number of computers being used in an internet café is counted every 30 minutes, starting at 9 am. Here are the results:

1 5 7 9 8 3 4 5 8 12 7 2 6 10 11 10

a Organise the data into a frequency table with four intervals ('1–3', '4–6', ...)

The result of one count shows that all the available computers are being used.

b What was the time of this count?

2. The number of books borrowed from a library is recorded each day. The results are:

61	50	35	62	74	53	47
55	67	56	45	62	42	41
66	55	59	64	46	48	69
68	58	50	54	30	49	70
79	65	59	56	57	46	60

a Organise the data into a frequency table with five intervals (30–39, 40–49 …)
b On how many days were less than 50 books borrowed?

3 The quality of two rival bus companies is compared by counting the number of late arrivals on 20 consecutive days. Here are the results:

Busy Buses	6	2	7	13	3	5	9	10	2	14
	8	1	4	11	3	17	4	8	4	6
Buses-R-Us	11	6	2	13	8	14	4	17	9	15
	10	18	2	7	16	1	7	13	20	18

a Organise the data for each company into a frequency table using four intervals ('1–5', '6–10', …)
b Which company offers the more reliable service? Justify your answer.

4 Tom's attendance at college is monitored over a 14 week period. Tom is timetabled to attend 15 lessons each week. The number of lessons Tom *misses* each week is shown in Table 3.5.

No. lessons missed	0–3	4–7	8–11	12–15
Frequency (No. of weeks) *f*	11	?	0	1

Table 3.5

a Find the value of the missing frequency ? in the table.
b Find the *greatest* possible number of classes Tom misses in total during the 14 weeks.

If Tom misses more than 30% of his classes in total then an absence letter is sent home to his parents.

c Is an absence letter sent home to Tom's parents? Show your working.
d Use the information above to complete Table 3.6 for the number of lessons Tom *attends* each week.

No. lessons attended	0–3	4–7	8–11	12–15
Frequency (No. of weeks) *f*				

Table 3.6

Intervals have class boundaries

Continuous data is always recorded to a particular level of accuracy. For example, the heights of plants may be recorded to the nearest centimetre. In this case, a height of 7.5 cm would be rounded to 8 cm.

The interval '8–10' contains all heights ranging from 7.5 cm to 10.5 cm.

7.5 and 10.5 are the *class boundaries* for the interval '8–10'.

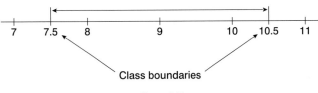

Class boundaries

Figure 3.11

Example

The table shows the heights of 33 plants recorded to the nearest cm.

Height (cm)	5–7	8–10	11–13	14–16	17–19
Frequency *f*	3	7	11	10	2

Table 3.7

Two plants with heights 7.7 cm and 13.5 cm must be added to the table. Complete the frequency table.

hint

Height is **continuous** data. All values from 4.5 cm to 19.5 cm are included in the table

Here's how...

The class boundaries (in cm) are: 4.5, 7.5, 10.5, 13.5, 16.5, 19.5

7.7 is between 7.5 and 10.5 and so belongs to the interval '8–10'

13.5 is one of the class boundaries. 13.5 is placed into the interval '14–16'

The new frequency table is shown below:

Height (cm)	5–7	8–10	11–13	14–16	17–19
Frequency f	3	8	11	11	2

Table 3.8

hint

Intervals include their lower boundaries.

checkpoint

$\Sigma f = 35$ as there are now 35 (= 33 + 2) plants being tallied.

Open-ended intervals can be used to avoid low frequencies

Intervals can be *combined* to avoid low frequencies. For example, the age of members of a golf club are:

Age (years)	10–19	20–29	30–39	40–49	50–59	60–69
Frequency f	2	8	29	45	41	52

Table 3.9

Compared to the other frequencies, the intervals '10–19' and '20–29' each have a low count. These intervals can be replaced by an *open-ended* interval, labelled 'under 30' by adding their frequencies:

Age (years)	under 30	30–39	40–49	50–59	60–69
frequency f	10	29	45	41	52

Table 3.10

Take care when dealing with ages. The interval '20–29' means from the day you are 20 up to and including the day **before** you are 30

Combining intervals has the disadvantage of losing some of the original information. It may not always be appropriate to use this technique.

> **Open-ended intervals should only appear at the beginning or end of a table.**

Now try these...

1 The heights of 27 students are recorded in centimetres, correct to one decimal place.

173.1 168.8 176.2 167.9 175.4 166.8 176.8 173.9 181.3
174.8 175.0 177.8 174.5 165.3 174.2 177.3 173.4 171.4
178.0 172.3 175.5 173.6 182.8 172.8 169.3 175.0 172.6

   a Organise the data into a grouped frequency table using 5 intervals '165–168', '169–172', ...
   b Which interval contains the most heights?

2 A computerised switchboard monitors the time taken for staff to answer customer complaints. Response times are recorded in seconds, to the nearest tenth of a second and are as follows:

2.5   10.6   8.8   19.8   22.2   4.7   9.4   30.4   6.7
30.2   26.8   35.6   25.7   23.8   14.7   9.0   11.3   10.7
38.2   15.8   33.4   3.6   13.4   1.3   39.7

   a Organise the data into a grouped frequency table using 4 intervals '1–10', '11–20', ...
   b Use the table to estimate the percentage of calls which took longer than 30 seconds to answer. Explain why your answer is an estimate.
   c Reorganise the data into five intervals '1–8 ', '9–16 ', ...
   d Which frequency table indicates greater efficiency of response times? Justify your answer.

3. In a 100 metre sprint competition, the first and second place finishers of each heat go through to the final. In total, four heats are held each with eight athletes. The table shows the finishing times (to the nearest 10th of a second) of all the runners in the first three heats.

Time (seconds)	Frequency *f*
9.8–9.9	6
10.0–10.1	9
10.2–10.3	4
10.4–10.5	5

Table 3.11

The class boundaries for the interval '9.8–9.9' are 9.75 and 9.95

   a Write down the class boundaries for the other intervals shown in the table.

The finishing times for the fourth heat are:

   9.87   9.92   9.96   10.07   10.13   10.26   10.38   10.42

   b Amend the frequency table above to include the finishing times for the fourth heat.
   c Which of these statements follow from the amended table?
      i   one quarter of the athletes completed their heat in under 9.95 seconds
      ii the qualifiers for the final all completed their heat in under 9.95 seconds
      iii nobody ran a heat time of 10.56 seconds.

4 Information on the weight of people registered at a health clinic is being prepared for the chief dietitian.

Weight (nearest kg)	40–49	50–59	60–79	80–99	100–119	120–129	130–139
Frequency *f*	1	1	12	18	9	0	2

Table 3.12

   a How many people are registered with the clinic?
   b Copy and complete the weights into the frequency table opposite:
   c Give two criticisms of Table 3.13.

Weight (nearest kg)	under 60 kg	60–99	100–119	120 kg or more
Frequency *f*				

Table 3.13

3.2.4 Checking your work for errors

When obtaining data it is important to make sure your results are accurate to avoid giving misleading information. Many checking techniques rely on common sense. For example, if tally marks are used to record sales, make sure the total number of tally marks adds up to the total number of items sold.

Now try this...

1 Jenny works as a shoe sales assistant in a small department store. One Saturday, Jenny sells 24 pairs of shoes. She records the size of each shoe sold as a tally mark. Here are her results.

Jenny looks at the table and realises she has made a mistake when counting the shoes.
   a Explain how Jenny knew she had made a mistake.
   b For which shoe size do you think Jenny made a mistake in her tallies? Explain your answer.

Shoe Size	No. shoes sold
8	ЖЖ ЖЖ II
8½	ЖЖ ЖЖ
9	ЖЖ I
9½	ЖЖ IIII
10	ЖЖ III
10½	IIII

Table 3.14

Working With Data

3.3.1 The three averages

An average is a single value which represents the data. For example, weather records show 'the average rainfall in London for July is 2 mm per day'. This does not necessarily mean that *exactly* 2 mm of rain fell on every day in July. It may have rained heavily on some days and been dry on others. The single value '2 mm per day' lies somewhere in between these two extremes and gives a rough idea of rainfall on a *typical* day in July.

The three types of average are the **Mode**, **Median** and **Mean**.

A mode is a value that occurs most often

Example

A shop sells four brands of computer:

> *Vixen* (V), *Chestnut* (C), *Obs-Elite* (O), *Hexium* (H)

Sales for one day are: C, H, V, O, C, H, C, V, C, O, C, O
State the mode.

Here's how...

It is helpful to arrange the sales in alphabetical order, crossing off each letter as you go:

> **C, C, C, C, C**, H, H, O, O, O, V, V
> ↑
> mode = most popular   **Answer:** The mode is *Chestnut*

checkpoint

12 items in original list.
12 items in ordered list.

To find the mode of quantitative data, it often helps to arrange the data into increasing (or *ascending*) order

Example

The administration section of a company employs eleven junior staff. Their weekly wages (in £) are:

> 170, 165, 190, 165, 180, 190, 160, 165, 185, 190, 165

Find the modal wage.

Here's how...

In ascending order, the wages are:

> 160, **165, 165, 165, 165**, 170, 180, 185, 190, 190, 190
> ↑
> mode = most common   **Answer:** The modal wage is £165

£165 is a reasonable answer as an average. Six staff (i.e over half) earn within £5 of the modal wage.

checkpoint

11 items in original list.
11 items in ordered list.

checkpoint

Makes sense – the mode is between the smallest and largest values.

Data may have two or more modes. For example, suppose the number of items of mail received at a house on nine mornings are: 2, **3, 3, 3**, 4, 4, **5, 5, 5**

The highlighted figures shows there are two modes 3 and 5. This data is **bi-modal**. Data with more than two modes is called **multi-modal**.

A modal value must occur at least twice. If all its values are different, the data does not have a mode.
For example, the data: 3, 5, 6, 7, 8, 10, 14 does not have a mode.

Now try these...

1 A newsagent sells the following copies of newspapers in one morning:

M	*F*	*M*	*D*	*D*	*E*	*M*	*F*
D	*M*	*D*	*F*	*D*	*M*	*E*	*M*

Key: *D* = Daily Blurb, *E* = Evening Star, *F* = Facts & Figures, *M* = The Moon

  a Arrange the data into alphabetical order.
  b Find the mode.

2 A particular type of savings bond costs £10. Cheques for bonds (in £) are received for the following amounts:

80	50	90	80	100	90	60
120	50	60	100	90	80	90

  a Arrange the data into ascending order.
  b Find the mode.
  c How many bonds in total are purchased?

3 Rachael's attendance at college is being monitored. The days on which she is late for morning classes over a term appears on the printout below:

M	W	Th	T	W	M	T
W	Th	T	W	M	T	W
W	Th	T	W	W	M	W
Th	T	W	M	T	W	Th

Key: M = Monday T = Tuesday W = Wednesday Th = Thursday F = Friday

  a Organise the days of the week into a frequency table.

Rachael has one day off from college per week for study leave.
  b Which day of the week do you think Rachael has for study leave? Explain your answer.

Rachael also has a part-time job one evening per week. Her shift finishes at midnight.
  c Which evening during the week do you think Rachael works? Explain your answer.

4 Diane is in charge of the bread section in a minimarket. At the start of each day, Diane stacks the shelves with equal numbers of wholemeal and white loaves of bread. At the end of each day, she notes the number of loaves of wholemeal bread left on the shelves.

The results for one week are:

Sat	Sun	Mon	Tues	Weds	Thurs	Fri
0	3	1	0	2	0	1

Table 3.15

  a What is the modal number of unsold wholemeal loaves?

The modal number of unsold loaves of white bread during the same week is three.
  b Which type of bread is more popular?

5 The number of calculators sold by a college bookshop is recorded over 12 days:

2	5	3	5	2	3	2	4	5	4	5	?

The sales figure for the last day ? has been lost.
  a Arrange the given data into ascending order.

The bookshop owner knows that the data for ***all*** the sales is bi-modal.
  b What is the missing sales figure? Explain your answer.

The median is the middle value in an ordered list

Example

The wages of the junior employees in the administration section are:

170, 165, 190, 165, 180, 190, 160, 165, 185, 190, 165

Find the median wage. Comment on the answer.

Here's how...

Arrange the wages in ascending order, *including any repeated values*:

160, 165, 165, 165, 165, 170, 180, 185, 190, 190, 190

Strike off outer pairs until you reach the middle of the list.

1̶6̶0̶, 1̶6̶5̶, 1̶6̶5̶, 1̶6̶5̶, 1̶6̶5̶, 170, 1̶8̶0̶, 1̶8̶5̶, 1̶9̶0̶, 1̶9̶0̶, 1̶9̶0̶
↑
median = middle value

Answer: The median is £170

The median is a suitable average. Everyone in the section earns an amount close to £170.

hint

Median:
1 Order the list.
2 Find the middle value.
3 If there are two middle values, add them and divide by 2 (see below).

checkpoint

Makes sense:the median is between the smallest and largest values.

The median was one of the wages listed (£170). This may not always be the case. Suppose the manager of the section earns £415 per week. Including his wage in the list gives:

1̶6̶0̶ 1̶6̶5̶ 1̶6̶5̶ 1̶6̶5̶ 1̶6̶5̶ 170 180 1̶8̶5̶ 1̶9̶0̶ 1̶9̶0̶ 1̶9̶0̶ 4̶1̶5̶
↑
median

The median is now the number halfway between 170 and 180 (i.e. halfway between the last pair of values to be crossed out).

$$median = \frac{170 + 180}{2} = \frac{350}{2} = 175$$

The median wage of the whole section is £175. Nobody in *Administration* earns the median wage. However, £175 is still reasonably close to what *most* people earn and so is a suitable average.

The manager earns considerably more than the junior employees. Including the manager's wage has only increased the median from £170 to £175. Since it is the middle value, *extreme values* do *not* significantly affect the median.

 Position of median = (Number of values + 1) ÷ 2

For example, with 11 junior employees: (No. values + 1) ÷ 2 = (11 + 1) ÷ 2
= 12 ÷ 2
= 6

i.e. the median occupies the 6th position in the ordered list of wages (check the example above).

With 12 values (i.e. including the manager), the rule gives the position of the median to be 6.5, which means halfway between the 6th and 7th value.

Be aware that this rule gives the *position* of the median in an ordered list, *not* the value of the median itself.

Now try these...

1 Find the median value for each of the following sets of data. In each case say whether you consider the median to be a good representative value, giving a reason.

   a Number of pints of milk delivered to a house each day during a week:

   | 2 | 2 | 3 | 2 | 2 | 3 | 4 |

   b Number of piglets in litters born on a pig farm:

   | 8 | 11 | 12 | 10 | 7 | 12 | 11 | 9 | 6 | 9 | 8 | 11 |

   c Heights of players on a rugby league team, measured to the nearest centimetre:

   | 180 | 195 | 184 | 179 | 190 | 175 | 184 | 193 | 187 | 184 | 186 | 192 | 183 |

   d Amounts collected from sponsors by a student who has completed a charity run:

   | 75p | 50p | £2.50 | 50p | £2 | 50p | £1.50 | 50p |

2 In order of decreasing merit, the grades awarded for an A level in French are   A, B, C, D, E, N, U. The following grades were achieved by a group of students:

   | A | B | D | A | E | D | B | C | E | U | B | C | N |

   a Arrange the grades in alphabetical order.   **b** Find the median grade.

   c Explain why there would be a problem in finding a median grade if there had been another student achieving grade D in the group.

3 The times (in seconds) taken by competitors in a 100 metre race were:

   | 10.5 | 10.8 | 11.3 | 10.4 | 10.3 | 11.5 | 10.7 | 11.5 |

   How many seconds quicker was the winning time than the median time?

4 The scores achieved in a test by a class of students are listed below:

   | 46 | 32 | 59 | 90 | 48 | 62 | 71 | 63 | 72 | 58 | 49 |
   | 52 | 84 | 64 | 66 | 70 | 82 | 12 | 74 | 75 | 34 | 82 |

   a How many students were tested?   **b** Find the median score.

   Two students were absent for the test and attempted it at a later date. One of these students scored 42 marks. When their results were included in the list the median remained the same.

   c Explain why the other student could **not** have scored 60 marks.

   d What is the smallest possible score of the other student?

The mean is the sum of the values divided by the number of values

Example

The wages of the junior employees in the administration section are:

170, 165, 190, 165, 180, 190, 160, 165, 185, 190, 165

Calculate the mean wage of the junior employees in administration.

Here's How...

Add up the wages:

Total wage = 170 + 165 + 190 + 165 + 180 + 190 + 160 + 165
+ 185 + 190 + 165

= £1925

$$\text{mean} = \frac{\text{sum of values}}{\text{number of values}} = \frac{1925}{11}$$

= £175   **Answer:** mean junior wage = £175

£175 is a representative value (although not one of the wages listed).

hint

Mean from a list:
1 Add the values.
2 Divide by the number of values.

checkpoint

Makes sense: the mean is between the smallest and largest values.

It is now quite easy to calculate the mean wage for the **whole** section, including the manager:

The total wage (in £) for the section = Total wage for juniors + manager's wage
$$= 1925 + 415$$
$$= 2340$$

i.e. mean wage $= \dfrac{\text{sum of values}}{\text{number of values}} = \dfrac{2340}{12} = £195$

All the junior staff earn less than £195 per week. The mean wage is *not* representative of the section. Because it uses all the data, the mean is affected by extreme values. ***Unusually large values will push the mean up, small values will pull it down***.

The table summarises the effect of introducing the manager's wage:

Juniors	Mode = £165	Median = £170	Mean = £175
Whole Section	Mode = £165	Median = £175	Mean = £195

Table 3.16

a mode	• **is a value which occurs most frequently (and at least twice)** • **is not significantly affected by extreme values** • **when present, is always one of the listed values** • **may not exist or there may be more than one (bi/multi-modal data)**
the median	• **middle value in an ordered list** • **is not significantly affected by extreme values** • **occupies position given by (No. values + 1) ÷ 2** • **is often one of the listed values**
the mean	• **mean = (Sum of values) ÷ No. values** • **uses all the data and so is affected by extreme values** • **is not usually one of the values in the list**

Now try these...

1 The daily distances (in km) travelled on a five-day walking tour are: 24, 18, 29, 25 and 19.
  Find the mean distance travelled per day.

2 Avril is making Christmas decorations to sell at a charity fair. One day she keeps a record of the number of decorations she makes each hour, with the following results:

        10   12   11   14   13   12

  a What is the mean number of decorations she has made per hour?
  b Avril plans to spend 30 more hours making decorations before the fair.
     Estimate how many more decorations she will make.

3 Imran and Paul are sitting end of term
  tests in Maths and Science at A level.
  Their results are shown in the table:

  Who achieved the higher mean score:
  a in Science   **b** overall?

Test	Maths	Biology	Chemistry	Physics
Imran	54%	63%	76%	47%
Paul	62%	48%	80%	52%

Table 3.17

4 Ten luxury assorted biscuits are weighed. The masses are given to the nearest gram:

        38g   42g   37g   39g   36g   40 g   38g   41g   38g   39g

  a Find the mean mass.
  b Estimate how many luxury assorted biscuits would be needed to give a total mass of 1 kilogram.

5 Tom delivers newspapers seven days a week. In one particular week, the mean number of newspapers delivered per day is 85.

How many newspapers does Tom deliver in that week?

6 A snooker player has a mean score of 52 from 9 games. In her tenth game she scores 147. What is the mean score for all ten games?

You may use a calculator for Q7–Q9 if you wish

7 The lengths of the tracks on a CD in minutes and seconds are:

   2:54   4:37   2:37   5:10   4:20   5:58   7:10   2:38   2:12   3:12   7:46

Calculate the mean length per track (hint: covert all times to seconds)
Give your answer in minutes and seconds, correct to the nearest second.

8 A swimming club has 37 male members and 23 female members. The mean age of the men is 24 years and the mean age of the women is 21 years. What is the mean age of all the members?

9 The temperature at a seaside resort is recorded at noon each day. The mean temperature for Monday to Friday in one week was 21.3°C. The mean temperature for Saturday and Sunday of that week was 18.5°C. What was the mean temperature for the whole week?

Which average is best?

Depending on the problem, one type of average might be more representative than another.

Example

The results of an A level group in a History test are:

   score (%)   3   8   42   57   59   60   63   64   65   67   73

a Calculate the:

   i median score   **ii** mean score   and comment on your answers.

b Explain why the mode is an unsuitable average for this data.

Here's how...

a i The scores are already in ascending order. Strike off outer pairs:

   ~~3~~ ~~8~~ ~~42~~ ~~57~~ ~~59~~ 60 ~~63~~ 64 ~~65~~ ~~67~~ ~~73~~

   ↑
   Median   **Answer:** Median = 60 %

> *hint*
> or… for 11 values the median occupies the $(11 + 1) \div 2 = $ 6th position.

ii Total score = 3 + 8 + … + 73
             = 561

There are 11 students in the group.

$$\text{mean} = \frac{\text{sum of values}}{\text{number of values}} = \frac{561}{11}$$

             = 51   **Answer:** Mean = 51 %

Comment: The scores suggest an able group, with over half the students scoring 60% or more.

median: 60% represents the group's ability. It is a suitable average to use.

mean: 51% indicates a weaker group. The mean is distorted by two low marks (3 and 8). It is *not* a suitable average for the group.

> *hint*
> Extreme values affect the mean but not the median.

b All the scores are different. The data does not have a mode.

In some situations, finding the mean may not even make sense. For example, Chinese takeaway food is often ordered using a system of numbers. The number '42' on a menu might stand for *Sweet & Sour Pork*.

Suppose a restaurant receives the following takeaway orders:

$$42 \quad 34 \quad 12 \quad 67 \quad 34 \quad 30 \quad 62 \quad 34 \quad 45$$

The mean of these numbers is: $\dfrac{42 + 34 + \ldots + 45}{9} = \dfrac{360}{9} = 40$

The number '40' may not appear on the menu and even if it does, it may refer to a dish that is rarely ordered. The mean is not a suitable average. The data appears to be numerical, but is actually qualitative, since each number describes a dish. The most appropriate average in this case is the mode (= 34).

Now try these...

1   Ten witnesses to a robbery are asked how many robbers took part in the crime. Their answers were:

$$5 \quad 4 \quad 5 \quad 3 \quad 4 \quad 4 \quad 5 \quad 5 \quad 5 \quad 3$$

   a   Find   i the median   ii the mode   iii the mean.
   b   Which of the three averages do you think is the most appropriate for the police to use in their investigation of the crime? Explain your answer.

2   A student is monitored for punctuality during a week of lectures. Each lecturer notes the number of minutes the student is late for class. The results are given below:

$$0 \quad 2 \quad 2 \quad 0 \quad 1 \quad 0 \quad 12 \quad 0 \quad 2 \quad 5 \quad 1 \quad 4 \quad 0 \quad 0 \quad 3 \quad 0$$

   a   Find   i the median   ii the mode   iii the mean.
   b   Which of these would the student prefer to use as the 'average'? Explain why.

3   An exhibition is held for a week. The attendance figures are:

Day	Mon	Tues	Wed	Thurs	Fri	Sat	Sun
Attendance	63	68	54	63	72	162	148

Table 3.18

   a   Find
      i the median   ii the mode   iii the mean.
   b   Which average gives the best indication of the popularity of the exhibition? Give reasons for your answer.

4   Jake collects bus numbers. He records the buses he sees on one Saturday. The results are:

$$90 \quad 281 \quad 222 \quad 90 \quad 43 \quad 120 \quad 67 \quad 97 \quad 281 \quad 90 \quad 71$$

   a   Find   i the median   ii the mode   iii the mean for the bus numbers.
   b   Which of the averages is the most representative for the buses he sees on this Saturday?
   c   Which average should Jake always use whenever he collects bus numbers?

5   The amounts won by 18 successive competitors in a television quiz show are:

£16 000	£8000	£1000	£8000	£8000	£4000
£4000	£32 000	£4000	£8000	£125 000	£1000
£8000	£16 000	£8000	£32 000	£4000	£8000

   a   Find   i the median   ii the mode   iii the mean amount won.
   b   Which of these averages do you think gives a typical prize?
   c   Which average would you advertise if you wanted to encourage people to take part in the quiz?

6   A tennis player has recorded some details of the matches he has played this year.

For each set of data **a**, **b** and **c** decide which of the three averages the tennis player would prefer to have reported in the local press. Show working to support your answers.

a No. of sets won	5	3	2	2	4	2	5
b No. of aces served	6	3	8	2	1	5	6
c No. of double faults served	0	1	3	4	4	5	4

Table 3.19

3.3.2 The range

Knowing the spread of the data is important. For example, some students prefer exams to be well spaced out to allow for extra revision (a high degree of spread of exam dates). Others like to get them over with quickly, preferring exams to be 'bunched' together (a low degree of spread of exam dates).

A simple measure of spread is the **range** – the difference between the highest and lowest value. A small range means the highest and lowest values are close together indicating a small degree of spread.

Example

The midday temperature (°C) in Rome is recorded over ten days in June:

23   24   25   27   **30**   26   22   **21**   26   29

Calculate the range.

Here's how...

The highest and lowest temperatures are highlighted.
the range = highest value − lowest value
= 30 − 21
= 9°C

Answer: Range = 9°C

> hint
> The range has the same units as the data.

Before finding the range, you must ensure all the values are given in the same units. For example, suppose the postage paid on six items is: 34p, 45p, 49p, 55p, 60p and £1.50

Most of the values are given in pence so we must convert £1.50 into pence:

£1.50 = 150 p   and the range   =   highest − lowest
=   150 − 34
=   116 pence

The range
• is the highest value − the lowest value
• can only be found if all the values are in the same units
• may be affected by extreme values

Now try these...

1   Find the range of the following data:
  a The marks awarded to an ice-skater by six judges in a competition 5.4   4.9   5.6   5.1   4.8   5.7
  b The masses of fish caught by an angler: 400 g   1.5 kg   650 g   2.3 kg   2.8 kg   780 g   1.2 kg

2   The temperature is recorded at a location on the coast and also at an inland location at three hourly intervals throughout a winter day. Here are the results:

Time	Midnight	3 am	6 am	9 am	Midday	3 pm	6 pm	9 pm
Coastal Location (°C)	−1	−3	−3	1	4	7	5	0
Inland Location (°C)	−2	−5	−4	1	5	8	4	−1

Table 3.20

  a Find the range of these temperatures for: **i** the coastal location **ii** the inland location.
  b Explain why it might be incorrect to say that the temperatures over the day varied more at the inland location than at the costal location.

3 An office manager has a choice of two routes
when travelling to work, one using a main
road and the other using minor roads. She
records the time taken to the nearest minute
for five journeys on each route.

Major Road	17	21	20	13	24
Minor Roads	20	18	21	18	19

Table 3.21

a Find the mean time and the range for each route.

b One morning the manager has an important meeting. She leaves for work just 20 minutes before
the meeting is due to start. Which route would you advise her to use? Explain your answer.

3.3.3 Analysing a frequency table

The rule: **position of median = (No. values + 1) ÷ 2** can be used to find the median value for data in a
frequency table. The **mode** is a value **with the highest frequency**.

Example

A test consists of four questions on mental arithmetic. The number of correct
answers achieved by 17 students is:

No. correct answers	0	1	2	3	4
No. students f	3	6	7	1	0

Table 3.22

Find the:
a median   **b** mode   **c** range for the number of correct answers.

hint

Understand the
information: six students
answered only one
question correctly.

Here's how...

a The median is the value in the 9th position.
From the table: 9th value
the first three values are all '0's: 0, 0, 0   ↓
the next six values are all '1's: 1, 1, 1, 1, 1, **1**   **Answer:** Median = 1

b The most common number of correct answers is 2.   **Answer:** Mode = 2

c The number of correct answers range from
0 to 3 inclusive. The range is $3 - 0 = 3$ **Answer:** Range = 3

hint

Take care: no-one
answered all four
questions correctly.

The rule: **mean = sum of values ÷ no. of values**
can be adapted to find the mean of a frequency
table.

For example, suppose 20 people attend a charity
dinner. Each person donates £5, £10 or £20 to the
charity.

Their contributions are given in the frequency table:

four people each gave
£5, contributing a total
of 5 × 4 = £20

five people each gave
£20, contributing a
total of 20 × 5 = £100

Donation x (£)	No. people f
5	4
10	11
20	5

Table 3.23

The amount donated in total is found by extending
the table to include an extra column 'xf':

Mean donation per person:

= total amount donated ÷ total no. people

$= \dfrac{\Sigma xf}{\Sigma f}$

$= 230 ÷ 20$

$= £11.50$

total No.
people

Donation x (£)	No. people f	xf
5	4	5 × 4 = 20
10	11	10 × 11 = 110
20	5	20 × 5 = 100
	$\Sigma f = 20$	$\Sigma xf = 230$

Table 3.24

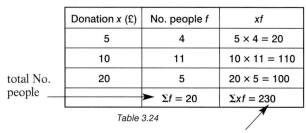

total amount donated

Σxf means the **sum of the values in the** xf **column**

For data in a frequency table: $mean = \dfrac{\Sigma xf}{\Sigma f}$

Now try these...

For each frequency table in questions **1** to **4**:
  a Find the: **i** modal value   **ii** median value   **iii** mean value   **iv** range   of x.
  b For each average give reasons why it may or may not be a good representative.

1 The number of damaged parcels received at a post office during August.

Number of damaged parcels x	0	1	2	3	4	5	6
Number of days f	6	10	0	11	2	1	1

Table 3.25

2 Number of tracks on the CDs recorded by the band *Pop Guns*.

Number of tracks x	9	10	11	12	13	14
Number of CDs f	1	0	5	6	1	2

Table 3.26

3 The number of occupants in 70 flats on an estate.

Number of occupants x	0	1	2	3	4	5	6
Number of flats f	2	27	19	11	7	3	1

Table 3.27

4 Age of children taken to see 'Santa' in a department store on a Saturday near Christmas.

Age of child x	1	2	3	4	5	6	7	8	9
Number of children f	0	14	23	21	35	33	6	2	1

Table 3.28

5 A die is thrown 100 times and the results are tallied.

Score x	1	2	3	4	5	6
Frequency f	10	15	11	16	20	28

Table 3.29

  a Find the **i** modal   **ii** median   **iii** mean score per throw.
  b Which of these averages would be most useful to you if you were guessing the next score?

6 A supermarket receives a large delivery of eggs packed in boxes of six. Fifty boxes are inspected and the number of cracked eggs counted. One of the frequencies is missing. Here are the results.

Number of cracked eggs x	0	1	2	3	4	5	6
Number of boxes f	39	?	3	0	1	0	1

Table 3.30

  a Find the value **?** of the missing frequency.
  b Find the **i** modal   **ii** median   **iii** mean number of cracked eggs per box.
  c Which of these averages could be used to estimate the total number of cracked eggs in the delivery.
  d Use your answer to part **c** to calculate an estimate for the total number of cracked eggs in the delivery if it consists of 2000 boxes.

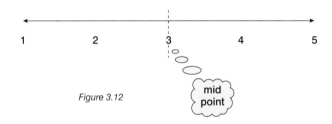

LEVEL 3 ## Averages can be estimated from a grouped frequency table

Score	No. students *f*
1–5	8
6–10	12
11–15	20
16–20	10

Table 3.31

When analysing grouped data, we can only find *estimates* for an average value. For example, suppose 50 students sit a Maths test marked out of 20. The scores are grouped into intervals as shown in Table 3.31:

Information is lost by grouping data. The interval '11–15' has the highest frequency and is called the **modal class**. However, we cannot tell which, if any, of the scores in this interval are the most common.

Similarly, although the table shows eight students scored between 1 and 5 marks, we do not know the precise value of any of these scores. In order to estimate the mean score of the group, we must assume each student achieves the **mid-point** score of each interval.

The mid-point of the interval '1 − 5' is the score half-way between the 1 and 5 (i.e. **3**)

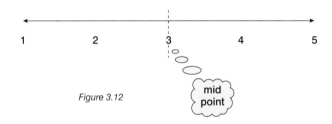

Figure 3.12

mid
point

We must assume these eight students each scored 3 marks. The table is extended as shown opposite:

Mean score $\approx \dfrac{\Sigma xf}{\Sigma f} = 560 \div 50$
$= 11.2$

11.2 is only an estimate for the mean. It is unlikely that each student achieves a score equal to the mid-value of the interval to which they belong. **Narrowing the width of each interval gives a better estimate for the mean** at the expense of having to carry out **more** calculations.

Score	Mid-point *x*	Frequency *f*	*x f*
1–5	3	8	3 × 8 = 24
6–10	8	12	8 × 12 = 96
11–15	13	20	13 × 20 = 260
16–20	18	10	18 × 10 = 180
		Σ*f* = 50	Σ*xf* = 560

Table 3.32

Note that intervals may be written without an upper end point. This is often the case for grouped **continuous** data. For example, Table 3.33 shows the time (in minutes) spent by a doctor consulting each of his patients during a five hour period.

Treatment time (mins)	5–	10–	15–	20–25
No. patients *f*	4	9	3	2

Table 3.33

The interval '5– ' appears to have no upper end-point. In this case, we take the upper value to be 10 (the starting value of the *next* interval in the table).

The mid-point of '5– ' is then calculated as: $\dfrac{5 + 10}{2} = \dfrac{15}{2} = 7.5$

The mid-points for the intervals are: 7.5, 12.5, 17.5 and 22.5. An estimate for the mean can then be found by using these mid-points as the *x* values and extending the table to calculate *xf*.

Now try these...

1 A typist has prepared an annual report. 30 paragraphs from the report are checked for spelling mistakes. The results have been grouped into intervals.

No. errors per paragraph	0–2	3–5	6–8	9–11
Frequency f	25	4	0	1

Table 3.34

  a Calculate an estimate for the mean number of errors per paragraph.

  The annual report consists of 135 paragraphs.

  b How many errors would the annual report be expected to contain? Give your answer to an appropriate degree of accuracy.

  Another selection of paragraphs from the same report estimates that the typist averages five spelling mistake in every nine paragraphs.

  c Which estimate (the first or the second) provides evidence of greater accuracy?

2 The manager of a printing firm employs 27 part-time staff. Their wages have been grouped into intervals:

Wage (£)	120–	140–	160–	180–200
No. employees f	11	6	8	2

Table 3.35

  a Calculate an estimate of the mean weekly wage.
  b State the modal class for weekly wages.

  The manager gives the part-timers a 5% pay increase.

  c Use your answer to **a** to estimate the new mean weekly wage for the part-timers.

3 At the start of their two year course, the heights (to the nearest centimetre) of a group of students are measured. Their heights are recorded again at the start of their second year. No one has joined or left the group during this time. The tables have *not* been labelled with the year in which the heights were measured.

Height (cm)	155–161	162–168	169–175	176–182	183–189
No. students f	2	3	7	5	2

Table 3.36

Height (cm)	155–160	161–166	167–172	173–178	179–184	185–190
No. students f	1	2	4	5	5	2

Table 3.37

  a Calculate an estimate for the mean of the heights in   **i** Table 3.36   **ii** Table 3.37
  b State the year (first year or second year) for each table, giving a reason for your answer.

4 The *MakingTracks* train company is required to publish its performance record. It produces two tables, each one showing the number of trains cancelled over the same 20 day period.

No. cancelled trains	0–4	5–9	10–14	15–21
No. days f	12	5	2	1

Table 3.38

No. cancelled trains	0–3	4–7	8–11	12–15	16–19
No. days f	5	9	3	2	1

Table 3.39

  a Without performing any calculations, state which table gives a more accurate estimate for the mean number of trains cancelled per day.

  If the mean number of trains cancelled per day is 6 or more then the company must pay a heavy fine.

  b Which table is the company most likely to publish? Show your working.

Presenting Findings

3.4.1 Drawing and analysing bar-charts

Bar-charts display frequency

A bar-chart consists of a series of rectangles, usually drawn vertically. The height of each rectangle represents the frequency of each value. Bar-charts can be used to illustrate qualitative data or discrete data.

Example

The table shows newspaper sales at a kiosk during a one hour period.

a Draw a bar-chart to illustrate the information.

b Find the percentage of sales that were *Small Print*.

Newspaper	No. copies sold *f*
Daily Reader	14
News & Views	12
Small Print	15
The Shield	9

Table 3.40

Here's how...

a The chart is shown below:

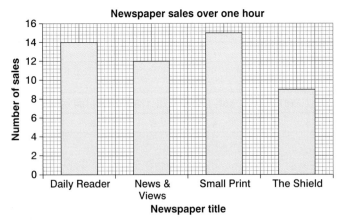

Figure 3.13

b The total number of sales is $14 + 12 + 15 + 9 = 50$
As a percentage of total sales, *Small Print* sales = $\frac{15}{50} \times 100 = 30\%$

Answer: *Small Print* sales = 30%

hint

The titles are placed on the horizontal axis. The vertical axis shows the number of copies of each title sold.

hint

One bar for each title is drawn. The first bar has height 14 corresponding to the sales figure for the *Daily Reader*.

checkpoint

or... add the heights of the bars to give the total frequency.

checkpoint

Makes sense: '15 in 50' is in the same proportion as '30 in 100', i.e. 30%.

When drawing a bar-chart, bear in mind the following points:

- Use a pencil for drawing the bars and a pen to clearly label the axes.
- Use a frequency scale which increases in 'convenient' steps (e.g. 2s, 5s, 10s, 20s, 50s etc..) avoiding 'awkward' scales such as steps of 3 or 8. Help with choosing a suitable scale is given in the next example.
- Make the chart as large as possible whilst keeping to a 'sensible' scale.
- The bars may be drawn horizontally.
- Give the chart a title.

Charts need sensible scales

Choosing a sensible scale is key to producing a chart which is easy to read. A good scale allows the whole range of values to be displayed whilst making the chart as large as possible. You may need to **break** a scale (not start from zero) in order to achieve this but if so mark this clearly as shown below.

Example

The table shows the number of parking tickets issued by a traffic warden per day over a period of 100 days.

No. tickets issued	5	6	7	8	9	10
No. days f	22	26	31	13	6	2

Table 3.41

Choose suitable scales to draw a bar-chart for this information.

Here's how...

Horizontal scale: The number of tickets per day vary from 5 to 10. Break the scale so that the values from 0 to 4 are ignored.

Vertical scale: The maximum frequency is just over 30. If one large square on the vertical axis is worth '5 days' then seven large squares are needed to reach 35.

Since each large square has 5 small squares up its side, each small square represents 1 day.

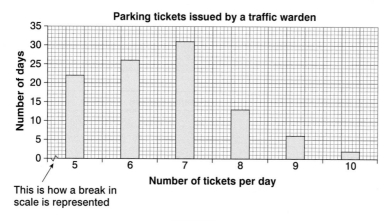

This is how a break in scale is represented

Figure 3.14

> **hint**
> By not breaking the scale, a large part of the bar-chart would consist of empty space.

> **hint**
> Work out the value of 1 small square.

> **checkpoint**
> '2 days' per large square would require 17 large squares (too many).
> '10 days' per large square would require 4 large squares (too few).

- **Use easy scales (e.g. 2s, 5s, 10s, 20s, 50s, etc.)**
- **A broken scale can give misleading information and so must be clearly labelled**
- **Avoid difficult scales**

Now try these...

1 The number of drinks bought from a machine during a lunch hour are shown in the table.

Type of drink	Black coffee	Coffee with milk	Black tea	Tea with milk	Hot chocolate
No. drinks f	24	39	7	42	28

Table 3.42

  a Choose a sensible frequency scale and draw a bar-chart to illustrate this information.
  b Which is the most popular drink?
  c Find the percentage of the drinks sold that were: i hot chocolate   ii coffee   iii tea.

2 The number of students enrolled for each part-time course at a college is counted and recorded in the table.

No. students enrolled on course	12	13	14	15	16	17	18	19	20
No. part-time courses f	2	1	3	7	0	4	10	8	2

Table 3.43

  a Choose a sensible scale and draw a bar-chart to show this information.
  b What is the modal number of students per part-time course?
  c How many part-time courses are there?

It is decided that any part-time course with less than 15 enrolments will not run.
  d How many of these courses will be cancelled?

3 A survey was carried out to assess the need for extra parking space at a college. Students with cars were asked to record how many times they were unable to find a parking space in the college car park during a term. The results of the survey are given in the table.

No. times student was unable to park	0	1	2	3	4	5	6
No. students f	62	51	34	27	18	6	12

Table 3.44

  a Choose sensible scales and draw a bar-chart to illustrate this data.
  b How many students took part in the survey?
  c Do you think extra car parking space is needed? Explain your answer.

4 In 1997, safety measures in *Hazard Land* were introduced to reduce the number of accidents in the work place. The number of accidents in *Hazard Land* each year after 1997 are shown in the table.

Year	1998	1999	2000
No. accidents f (1000s)	500	495	490

Table3.45: Accidents 1998–2000

  a Draw a bar-chart for this information using:
    i a broken vertical scale and large squares labelled 0, 490, 492 etc...
    ii a complete vertical scale with large squares labelled 0, 100, 200 etc...

A Health & Safety officer wants to show colleagues that the new safety measures have not significantly reduced the number of accidents in the work place.
  b Which of your bar-charts is the officer most likely to choose? Justify your answer.

Bar-charts and histograms can be used to show grouped data

For grouped data, one bar is drawn for each interval. If the data is discrete and covers a fairly small range, the bars can be separated by gaps. For example, the number of calls made on mobile phones over a weekend by a group of students is shown in Table 3.53:

No. calls made	No. students *f*
1–5	5
6–10	8
11–15	12
16–20	3

Table 3.46

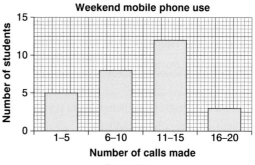

Figure 3.15

A bar-chart for this information has been drawn.
The first bar covers the interval '1–5 '
The second bar covers the interval '6–10 '
and so on…

In the example above, gaps between the bars re-enforce the fact that the data is discrete (i.e. the data can only take certain values) and covers a small range. When discrete data covers a range large enough to make the gap between intervals insignificant, it is acceptable to draw the bars so that they are touching.

Continuous data, on the other hand, can take *any* value within a given range. When displaying continuous data, the bars should be joined together at the class boundaries to form a **histogram**. If the intervals are of equal width, the height of each bar represents frequency.

The table shows the mass of luggage items checked in at an airport departure desk during a one hour period
A histogram has been drawn for the data. The bars meet at the class boundaries 0.5, 5.5, 10.5, etc.

Luggage mass (kg)	No. items *f*
1–5	74
6–10	83
11–15	53
16–20	27
21–25	18

Table 3.47

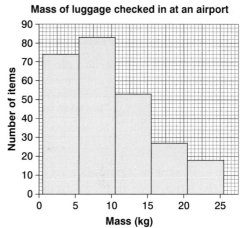

Figure 3.16

A bar-chart	•	is used to illustrate qualitative and discrete data (grouped or un-grouped)
	•	usually has gaps between the bars
A histogram	•	is used to illustrate continuous data
	•	must not have gaps between its bars
	•	has bars drawn at the class boundaries

Now try these…

1 Applicants for a job are given a literacy test. The number of spelling mistakes made by each applicant is recorded.
The results are shown in the table:

No. spelling mistakes	1–5	6–10	11–15	16–20	21–25
No. applicants *f*	16	26	17	9	2

Table 3.48

a Draw a bar-chart for this data.
b How many applicants made 10 errors or less?
c What percentage of the applicants made more than 10 errors?

2 A batch of plants is delivered to a garden centre. The height of each plant is measured to the nearest cm.

Height (nearest cm)	20–24	25–29	30–34	35–39	40–44	45–49
No. plants f	18	31	19	34	28	10

Table 3.49

a Draw a histogram showing these results.

The plants were advertised to have a minimum height of 30 cm to the nearest cm.
b What percentage of the plants fail to meet this description?

3 A supermarket carries out a survey to find out how long customers have to wait in the queues at check-outs before they are served. The results recorded one day at a particular check-out are given in the table.

Time (minutes)	0–	2–	4–	6–	8–	10–(12)
No. people f	85	92	64	56	32	21

Table 3.50

a Use a sensible scale to draw a histogram to illustrate this data (the class boundaries are 0, 2, 4, etc.)
b How many customers were served at this check-out during the day?

The supermarket has been set a target of serving 70% of customers in under six minutes.
c Is this particular check-out achieving this target? Justify your answer.

4 The time taken to write a program by each student on a computing course is measured to the nearest minute. The results are summarised in the table.

Time taken (nearest minute)	11–20	21–30	31–40	41–50
No. students f	13	22	13	8

Table 3.51

a Draw a histogram showing this information.
b Write down the modal group.
c Estimate the percentage of students who took longer than half an hour to complete the task.
d Explain why your answer to part **c** is an estimate.

Comparative and component bar-charts show how two quantities compare

Example

The comparative bar-chart shows UK employment figures (in millions) for males and females aged 16 or over in a country.

hint

Comparative bar-charts must always have a key.

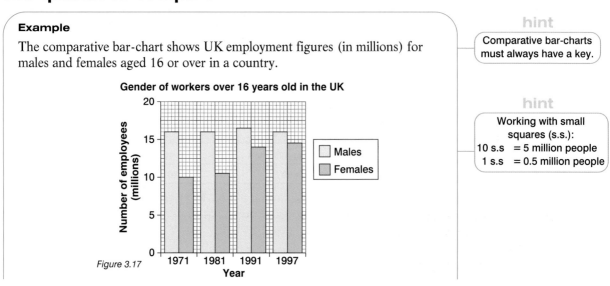

Figure 3.17

hint

Working with small squares (s.s.):
10 s.s = 5 million people
1 s.s = 0.5 million people

a How many:
  i males were employed in 1991
  ii adults were employed in 1981?

A government official claims that female employment is increasing as a percentage of the total work force.

b Is the official correct? Justify your answer.

Here's how...

a Use the key to select the correct bar. Reading from the frequency scale:
  i The number of males employed in 1991 = 16.5 million

  Answer: 16.5 million males

  ii In 1981 there were 26.5 million adults employed.

  Answer: 26.5 million adults

b The claim is **correct**. Female employment is increasing (the bars are getting taller) whilst male employment remains fairly constant. Female employment is increasing as a percentage of the total work force.

A comparative bar-chart is useful for comparing quantities. It is less convenient for comparing overall totals. A *component* bar-chart is drawn by stacking one bar *on top of* another. It is then easy to see totals and how each bar is made up.

Example

The component bar-chart shows the number (in tens of thousands) of male and female secondary school teachers in England and Wales.

a How is the total number of teachers changing?

b Find the number of male teachers in 1988.

c An spokesman states that 'from 1994 onwards, women have outnumbered men in the teaching profession'. Is the spokesman correct? Give a reason.

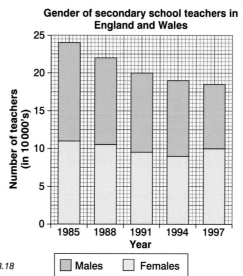

Figure 3.18   Males ☐ Females

hint

Working with small squares (s.s.):
10 s.s = 50 000 teachers
1 s.s = 5000 teachers

hint

Total number of teachers in 1994
= 200 000 – 2 small squares
= 200 000 – 10 000
= 190 000

hint

Component bar-charts must always have a key.

Here's how...

a For each year shown, the *total* heights of the bars is decreasing.

  Answer: The number of teachers is decreasing

b The unit of frequency is 10 000. Reading from the 1988 bar:

Total no. teachers	$= 22 \times 10\,000$	$= 220\,000$
No. female teachers	$= 10.5 \times 10\,000$	$= 105\,000$
No. male teachers	$= 220\,000 - 105\,000$	$= 115\,000$

  Answer: 115 000 male teachers

c There are more women than men teaching in 1994 and 1997 (compare the female and male bars for each year). The spokesman *appears* to be correct. However, no information on teacher numbers is given for 1995 and 1996. More information is needed.

Now try these...

1 The number of male and female members in each sports club organised by a student union is given in Table 3.52:

Draw a chart that illustrates the difference in the popularity of these sports between males and females.

Sports club	No. of male members	No. female members
Swimming	13	7
Cycling	15	10
Tennis	9	11
Badminton	8	14
Climbing	7	4
Windsurfing	10	6

Table 3.52

2 The table shows the reasons given by students on a course for absences during a term.

Draw one chart that shows the total number of absences due to each cause and the way in which each total is made up from male and female absences.

Reason	Male students	Female students
Cold or flu	24	20
Other illness	14	26
Medical appointment	10	18
Interview	8	10
Holiday	6	4
Other	5	7

Table 3.53

3 Table 3.54 shows the medals won by some of the European countries that took part in the Summer 2000 Olympic Games in Sydney.

a Use this data to draw: i a component bar-chart
 ii a comparative bar-chart.
b Which of the two charts do you think shows the differences between the performance of these countries most clearly? Give a reason for your answer.
c Write a brief summary of the main differences between the results of these countries.

Country	Gold	Silver	Bronze
Britain	11	10	7
France	13	14	11
Germany	14	17	26
Italy	13	8	13
Netherlands	12	9	4
Spain	3	3	5

Table 3.54

4 The number of students enrolled for some of the courses offered by a college are shown in the bar-chart.

a Which course is most popular
 i overall
 ii with males
 iii with females?
b On which courses are there more females than males?
c What percentage of the students enrolled for Engineering are male?
d Calculate the percentage of all the students shown that enrolled for Computing.

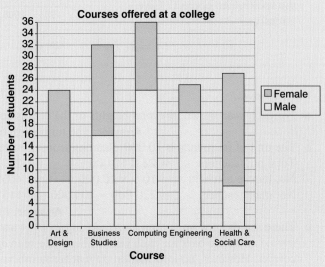

Figure 3.19

5 The bar-chart shows the grades awarded to students for 3 assignments completed on a computing course.

   a How many students failed the first assignment?

   b Find the total number of distinctions awarded for assignments.

   c Which assignment do you think was the easiest? Explain your answer.

   d What percentage of first assignments were not failed?

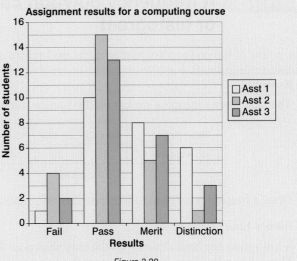

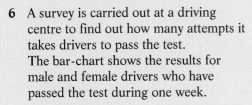

Figure 3.20

6 A survey is carried out at a driving centre to find out how many attempts it takes drivers to pass the test.
The bar-chart shows the results for male and female drivers who have passed the test during one week.

   a What percentage of males passed the test at their first attempt?

   b What percentage of females passed the test at the third attempt?

   c Compare the performance of the males with that of the females.

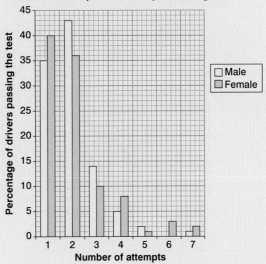

Figure 3.21

7. The bar-chart shows the performance of the Mayfield United football team in the matches it played last season.

   a Mayfield United played each of the other teams in their league twice.
   How many teams including Mayfield were in the league?

   b Compare Mayfield's performance in their home games with their performance in away games.

   c Mayfield were awarded two points for each win at home, three points for each win away and one point for each draw. Calculate the total number of points Mayfield had at the end of the season.

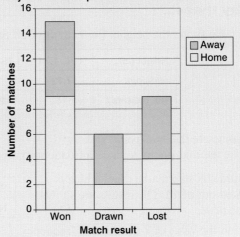

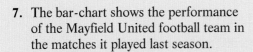

Figure 3.22

LEVEL 3 A frequency polygon can be drawn on top of a bar-chart or histogram

The points of the polygon are plotted at the mid-points of the top of each bar.

Example

The table shows the mass (in kg) of 100 female students.

Mass (kg)	40–(50)	50–(60)	60–(70)	70–(80)
No. students f	11	18	43	28

Table 3.55

Draw a frequency polygon to illustrate this information.

Here's how...

A histogram has been drawn with the bars shown as dotted lines. The polygon has been drawn on top of the bars.

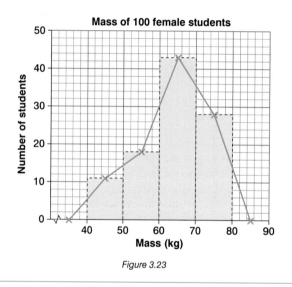

Figure 3.23

hint

The interval '40–(50)' includes all numbers in the range 40 to 50 but not including 50 itself. We take the mid-point of '40–(50)' to be 45.

hint

Complete the polygon by drawing lines to points 35 and 85 on the horizontal axis.

Sometimes the polygon is constructed without first drawing the bars. In this case, the frequency values must be plotted against the mid-points of the intervals.

Now try these...

1 The table shows the heights (in metres) of 60 patients recorded to two decimal places.

Height (m)	1.70–1.74	1.75–1.79	1.80–1.84	1.85–1.89	1.90–1.94
mid-point	1.72				
No. patients f	15	27	11	5	2

Table 3.56

   a Complete the 'mid-point' row.
   b Use the mid-points to draw a frequency polygon for this information

2 96 students sit a maths exam which is marked out of 40. The results are shown in the table.

Score	5–10	11–16	17–22	23–28	29–34	35–40
No. students f	7	11	24	21	?	5

Table 3.57

   a Find the value of ? the missing frequency.
   b Draw a frequency polygon for this information.

Two polygons can be used to compare two sets of data

Example

The frequency polygons show the mass (in kg) of parcels handled by two post offices on one particular day.

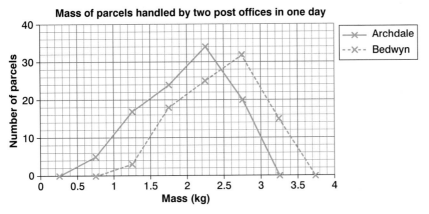

Figure 3.24

Use the chart to state which post office:
a generally handled heavier parcels,
b handled the most parcels.

Here's how...

a Both polygons have a similar shape but compared to Archdale, the polygon for Bedwyn is about 1 kg further to the right. The parcels handled in Bedwyn were generally 1 kg heavier than those handled in Archdale.

Answer: The post office in Bedwyn

b Find the total frequency for each polygon:
Archdale: Total no. parcels $= 5 + 17 + 24 + 34 + 20 = 100$
Bedwyn: Total no. parcels $= 3 + 18 + 25 + 32 + 15 = 93$
Archdale handled seven more parcel than Bedwyn's post office.

Answer: The post office in Archdale

hint

We could also say 'the parcels at Bedwyn are *on average* 1 kg heavier than those at Archdale'.

Now try these...

1 The midday temperature in London and Paris is recorded during August. The results are as shown.

a Which city generally had the higher midday temperatures?

b What was the approximate midday temperature difference between London and Paris in August?

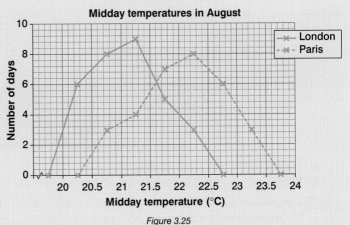

Figure 3.25

2   Blood pressure readings taken immediately after exercise tend to be higher than readings before exercise. A group of students measure their blood pressure (in mmHg). They exercise for five minutes and then measure their blood pressure again. The results are recorded to the nearest whole mmHg.:

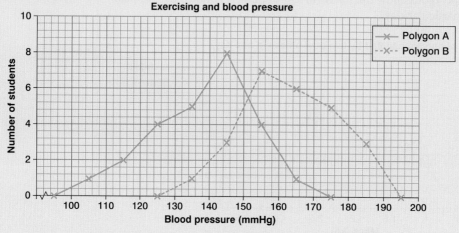

Figure 3.26

a   Which polygon shows the blood pressure of the group after exercising?
b   Use polygon A to find the number of students in the group. Check your answer using polygon B.

Inderjit was the student with the highest blood pressure before exercising.
c   What was Inderjit's lowest possible blood pressure reading before exercising?

After exercising, Inderjit's blood pressure was the fourth highest in the group.
d   What is the greatest possible difference between the two readings for Inderjit?

3   Two groups of students sat the same two hour Psychology exam. Group A sat the exam in the morning, Group B in the afternoon. There was no contact between the two groups and all candidates finished early.

In each group, students recorded how many minutes early they finished. The results are shown below:

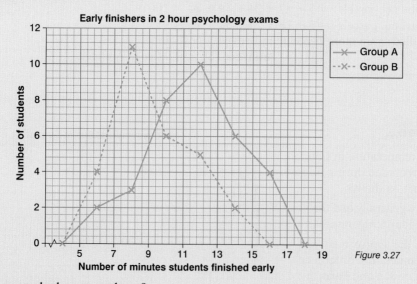

Figure 3.27

a   Which group had more students?
b   Which group generally took less time to complete the exam?

The exams are marked and the results show both groups did equally well.
c   Do the frequency polygons support the view that students work more productively in morning exams than in afternoon exams? Discuss your answer fully.

3.4.2 Drawing and analysing pie-charts

A **pie-chart** is a good way of illustrating data that has been split into categories. A circle is drawn and divided into **sectors**, each sector forming a slice of the pie. The size of each sector is proportional to the number of items in the category it represents.

The angles in a pie-chart represent frequency

 A pie-chart must have a title
The sectors of a pie-chart must be clearly labelled (possibly by using a key and shading)

Example

The table shows the destination of 30 people going on holiday:

Destination	Australia	Bahamas	Canada	Denmark	Egypt
No. people f	10	5	7	2	6

Table 3.58

Draw a pie-chart to illustrate this information.

Here's how...

Since there are a total of 30 people and a circle has 360°, each person is represented by $360° \div 30 = 12°$
The angles for each sector are worked out as follows:

Destination	f	Angle
Australia	10	$10 \times 12° = 120°$
Bahamas	5	$5 \times 12° = 60°$
Canada	7	$7 \times 12° = 84°$
Denmark	2	$2 \times 12° = 24°$
Egypt	6	$6 \times 12° = 72°$
	$\Sigma f = 30$	$\Sigma = 360°$

Table 3.59

hint
We could also use fractions:
Australia $= \frac{10}{30} \times 360° = 120°$

checkpoint
The sum of angles must equal 360°.

One sector for each destination is drawn as shown below:

Holiday destinations of 30 people

Figure 3.28

hint
Label the sectors with the destinations.

Now try these...

1 40 students were asked to name the sport they most enjoyed playing. Here are the results:

Sport	Golf	Tennis	Badminton	Football	Snooker	Other
No. students	3	7	6	10	8	6

Table 3.60

Draw a pie-chart for this information.

2 45 students sit a History test marked out of 20. The results are:

Score	1–5	6–10	11–15	16–20
No. students f	5	?	21	8

Table 3.61

a Find the value of ? the missing frequency.
b Draw the pie-chart for this information.

3 Attendance figures at a city's top tourist sites for one week are:

Attraction	Madame Two Swords	The Museum of Science	The Torture Chamber	Shuttle Park Zoo	Archaeological Dig Site
No. visitors (in 1000s) f	45	78	57	68	22

Table 3.62

Draw a pie-chart for this information (work to the nearest whole degree where appropriate).

4 The social backgrounds of students who entered a university in 2000 are:

Social background	Professional	Intermediate	Skilled	Unskilled	Unknown
% of Total admissions	25	40	15	12.5	?
Angle	$\frac{25}{100} \times 360° = 90°$				

Table 3.63

a Find the value ? for the 'Unknown' percentage.   **b** Copy and complete the table.
c Draw a pie-chart for this information.

5 The component bar-chart shows the sales of new and second-hand cars made from a garage over a two month period.

Figure 3.29

Draw a pie-chart showing sales of each make of car as a proportion of:
a total sales   **b** second-hand sales.

A pie-chart shows the proportion of each 'part' out of the 'whole'

Some proportions are easily found just by looking at the pie-chart.

Pie-charts also allow 'parts' to be compared with each other

Types of chocolates in a box

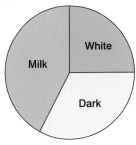

Figure 3.30

Train service one morning at a station

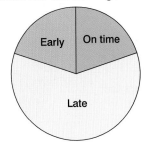

Figure 3.31

For this box of chocolates:
$\frac{1}{4}$ (= 25%) are white chocolates.
$\frac{1}{3}$ (≈ 33%) are dark chocolates.
Under half (= 50%) are milk chocolates.

On this morning approximately:
Equal proportions of trains arrived 'early' and 'on time'
Three times as many trains arrived late as arrived early.

It is sometimes necessary to find how many degrees each item is 'worth' in the pie-chart.

Example

45 compact discs are categorised as *Classical*, *Pop*, *Jazz* and *Rock*.

CD collection

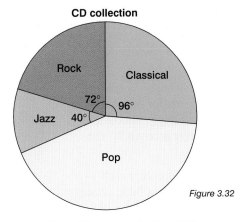

Figure 3.32

Find the number of: **a** *Jazz* CDs   **b** *Pop* CDs

Here's how...

a 45 CDs are represented by 360°
so 1 CD is represented by 360° ÷ 45 = 8°
The *Jazz* sector has angle 40°. The number of *Jazz* CDs is 40° ÷ 8° = 5

Answer: There are 5 *Jazz* CDs

b The *Pop* sector angle = 360° − (96° + 40° + 72°)
= 360° − 208°
= 152°
The number of *Pop* CDs = 152° ÷ 8° = 19

Answer: There are 19 *Pop* CDs

> **hint**
> Remember: 360° always represents the total frequency.

> **checkpoint**
> ...or use fractions:
> Jazz = $\frac{40}{360}$ × 45 = 5

> **hint**
> Use the sum of all angles = 360°.

Information about one sector can be used to find out information about other sectors or the whole.

Example

A group of students write down how many hours of television per channel they watch at home in a week. Their results are combined into a pie-chart.

If the group watched BBC1 for a total of 75 hours calculate:

a the total number of hours the group watched television

b the number of hours they watched Channel 4.

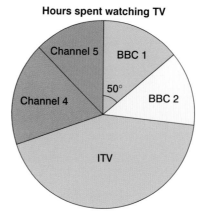

Hours spent watching TV

Figure 3.33

Here's how...

a The total time is represented by 360°.
Using proportion: 50° represents 75 hours (BBC1 sector)
so   1° represents 75 ÷ 50 = 1.5 hours
i.e.   360° represents 360 × 1.5 = 540 hours

                  Answer: Total viewing time = 540 hours

b By measurement, the angle for Channel 4 is 64°
From part **a**, 1° represents 1.5 hours
so   64° represents 64 × 1.5 = 96 hours

          Answer: Channel 4 was watched for 96 hours

hint

Use proportion: find the value of 1°.

checkpoint

Makes sense: the Channel 4 slice is larger than the BBC1 slice

Now try these...

1 Sherlock Holmes is looking at a pie-chart showing the various methods used to commit murder.

Through experience of solving crimes, he knows: Strangulation is the most common method of murder. Poisoning accounts for about 25% of all murders and is twice as common as death by shooting. Shooting is 3 times as common as death by drowning. Stabbings account for the remaining cases.

Without measuring its angles, decide which sector represents each murder technique. You may assume that the pie-chart has been drawn accurately.

Murder techniques

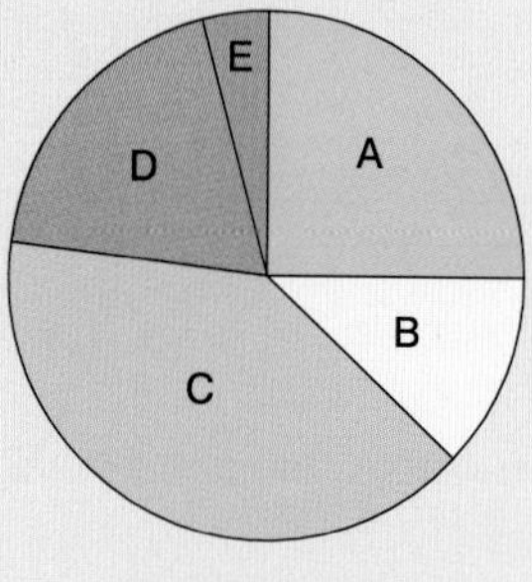

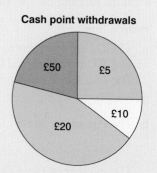

Figure 3.34

2 The amounts withdrawn from a cash point during a particular day are shown on the pie-chart which has been drawn accurately.

a Explain why pie-charts drawn using each table would have the same angle for £5.

b Without measuring any angles, state which of the tables the pie-chart represents. Give a reason.

Cash point withdrawals

Amount (£)	5	10	20	50
Frequency *f*	15	8	30	7

Table 3.64

Amount (£)	5	10	20	50
Frequency *f*	10	5	17	8

Table 3.65

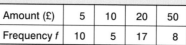

Figure 3.35

3 A car parking meter charges the following amounts:

30 p (up to 1 hour), 50 p (between 1 and 2 hours), 80 p (between 2 and 3 hours).

The times for which cars were parked during a particular day are shown on the accurately drawn pie-chart.

432 cars in total used the car park on that day. Find, by measuring angles, the:

a number of cars that were parked for between 1 and $1\frac{1}{2}$ hours,

b total amount of money received by the meter over the day.

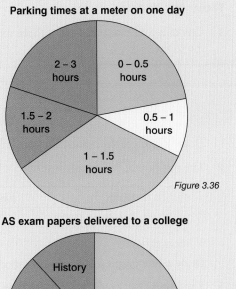

Parking times at a meter on one day

Figure 3.36

4 The pie-chart shows the proportion of AS exam papers in five subjects delivered to a college at the start of exam week. You may assume the exact number of papers required for each subject are sent and that the pie-chart is accurately drawn.

360 Sociology papers were delivered to the college. By measuring angles determine which of these statements must be true:

A 240 Biology papers were delivered.
B 2700 exam papers were delivered in total.
C 2700 students are scheduled to sit the exams.
D More than 1400 students are entered for English.

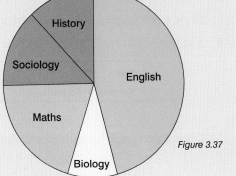

AS exam papers delivered to a college

Figure 3.37

3.4.3 Drawing and analysing line graphs

A *line graph* is a series of points plotted from given data and joined with straight lines. Line graphs are used to show trends and patterns in the data.

Drawing a line graph

Example

The table shows the monthly rainfall in a city from January to June:

Month	Jan	Feb	Mar	Apr	May	June
Rainfall (mm)	70	104	82	84	42	20

Table 3.66

Draw a line graph to illustrate this information.

Here's how...

Plot the month on the horizontal axis and the rainfall on the vertical axis. Join consecutive points with straight lines.

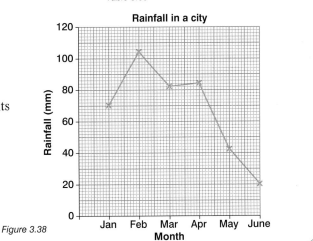

Figure 3.38

Now try these...

1 Gas bills arrive at a flat every three months. Each amount has been recorded to the nearest £ in the table.

Month	March	June	September	December
Amount (£)	45	27	17	35

Table 3.67

Draw a line graph to illustrate this information.

2 The fox population in a forest is:

Month	June	July	August	September	October
Population (100s)	5.0	7.5	3.1	8.7	4.6

Table 3.68

Draw a line graph for this information.

3 The number of employed 16–24 year olds in the UK were:

Year	1969	1979	1989	1999
No. employed (millions)	4.82	5.94	5.63	4.05

Table 3.69

Draw a line graph for this information (hint: break the vertical scale).

Analysing a line graph

The line graph shows sales of a brand of sun cream at a UK department store from March to July of a particular year. The number of tubes of cream sold is shown on the vertical scale.

Total number of tubes sold $= 22 + 32 + 73 + 85 + 97$
$= 309$ tubes

The sales figure for August can be ***estimated*** by extending the line to point A (see the dotted line on the graph). The estimated sales figure for August is 116 tubes. The estimate will only be reasonable if the sales pattern from previous months continues into August. In fact, August is the most popular time for taking holidays and so sun cream sales may well decline in August!

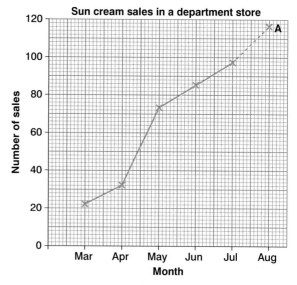

Figure 3.39

When interpreting a line graph:
- Look carefully at the units on each scale
- Look for a break in either scale
- Unknown values found using the lines are only estimates.

A line graph may have more than one line. A key or clear labelling of each line must be used to distinguish between the lines. Comparisons of the lines can then be made.

Suppose the department store in the previous example also sells umbrellas. The sales of umbrellas over the same time period are included on the sun cream line graph (see Figure 3.40).

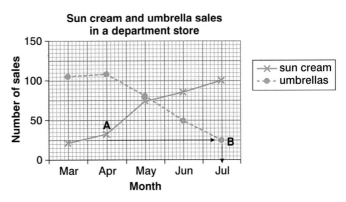

Figure 3.40

Example

a Using Figure 3.42, in which month:
  i were the number of umbrella and sun cream sales almost equal?
  ii was the number of umbrellas sold roughly equal to the number of sun cream tubes sold in April?
b Give a possible explanation for the shape of the graph.

Here's how...

a i Look for a month in which a cross (✗) and a dot (●) are closest together. **Answer:** May
  ii 32 tubes of sun cream were bought in April (see point A). Read across from point A to the umbrella graph (see point B). **Answer:** July

b The summer months tend to bring hot, dry weather. People tend to buy sun cream and not umbrellas in the summer months.

hint
Line graphs with two or more lines must use a key.

hint
The point at which the lines cross has no significance.

Now try these...

1 Orders placed at a turkey farm from December 1998 to May 1999 are shown on the line graph.

  a Find the total number of turkeys ordered.
  b Approximately what percentage of orders were placed in December?
  c Give an explanation for the shape of the line graph.

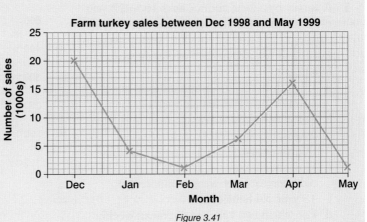

Figure 3.41

2 The line graph shows the air temperature close to the surface of a village pond which has frozen over. Readings are taken every half hour.

  a Use the line graph to estimate the range of temperatures:

    i between 7 am and 9.30 am

    ii 8:15 am and 11:45

  b Explain why your answer to **a** might only be an estimate.

For safety reasons, villagers are not allowed to skate on the pond when the temperature is 0°C or above

  c Use the line graph to estimate the latest time villagers should stop skating.

  d Why should your answer to part **c** be treated with caution?

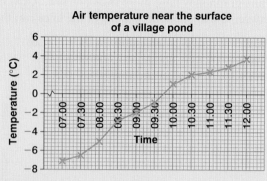

Figure 3.42

3 The *lowest* day time temperatures in Amsterdam over a six day period are shown on a line graph.

  a What is the highest recorded temperature?

  b Which two dates had the same recorded temperature?

  c Which of these statements follows from the graph?

    A The coldest day was the 18th June.

    B The hottest day was the 11th June.

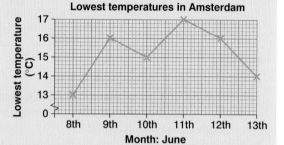

Figure 3.43

4 Tom is drawing a line graph for the amount charged by a car parking meter.

The meter charges the following rates:

up to 1 hour = 50p

20p for each extra hour (or *part* of an hour)

Points for various lengths of stay have been plotted on the line graph

To turn the chart into a line graph, Tom joins the points with straight lines.

  a Explain why it is incorrect to join points A and B together.

  b Draw a correct line graph for charges made.

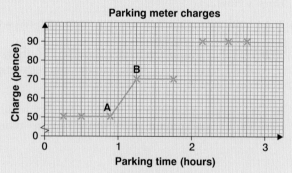

Figure 3.44

5 A survey by a local council asks 16 to 22 year old males and females to declare their weekly earnings. The average (mean) weekly wage for males and females at each age is shown on the line graph.

  a What does the line graph indicate about average male and female wages?

  b For which age are average male and female wages:

    i furthest apart?   **ii** closest together?

  c How much more on average does an 18 year old male earn per week compared to 18 year old female?

  d For which two ages are the mean weekly wage differences equal?

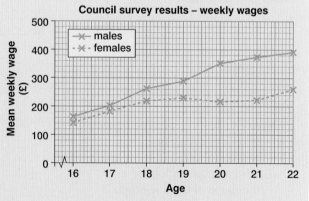

Figure 3.45

6 The number of births and deaths in a country from 1993–2000 is shown on the line graph.

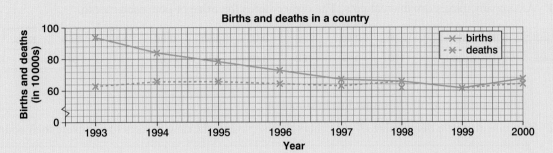

Figure 3.46

  a Briefly describe what each line shows.
  b Which of these statements does the graph prove?
    i The greatest difference between births and deaths occurred in 1993.
    ii The highest number of births and deaths occurred in the same year.
    iii The population of the country did not change from 1998 to 1999.

3.4.4 Drawing and analysing scatter graphs

A **scatter graph** is used to find a possible relationship between two variables such as height and weight. In most cases it is reasonable to expect taller people to weigh more than shorter people. If this is the case, a scatter graph of height against weight would show this connection.

Drawing a scatter graph

Suppose ten students travel to college by bus. Their distances (in miles to one decimal place) and journey times (to the nearest minute) into college are:

Distance (miles)	3.2	3.4	4.1	5.0	4.5	5.6	4.9	4.3	3.8	5.5
Times (minutes)	13	15	18	22	20	23	21	18	16	24

Table 3.70

The first student travels 3.2 miles in 13 minutes. This is marked as a single cross (✗) on the graph (see **[1]**). The scatter graph is drawn below. For convenience, **both** scales have been broken.

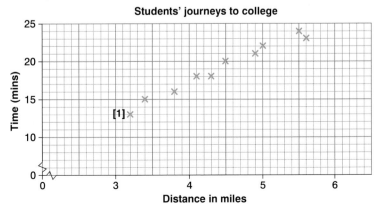

Figure 3.47

Now try these...

1 A company makes ornamental pots. The owner varies the price of pots each month.

Price (£)	2.00	2.50	2.45	2.90	3.50	3.95	4.65	5.55	5.95
No. sales	170	145	162	143	124	112	91	67	52

Table 3.71

Draw a scatter graph to illustrate this information.

2 Ten milkmen give the number of customers they serve and the time it takes to complete their round.

No. customers	45	59	33	48	30	65	67	50	78	39
Time (hours)	$2\frac{1}{2}$	3	$1\frac{1}{2}$	$2\frac{1}{4}$	1	$3\frac{1}{2}$	$3\frac{3}{4}$	$2\frac{1}{2}$	$4\frac{1}{2}$	2

Table 3.72

a Convert the times into decimal numbers.
b Draw a scatter graph to illustrate the information (hint: break the 'customer' scale).

Analysing a scatter graph

Common sense says that as journey distances increase, journey times should also increase. This is shown in the scatter graph in the previous example. The points lie roughly on a straight line sloping upwards from left to right. This is known as **positive correlation**.

Examples of scatter graphs with **negative** and **no** correlation are shown below:

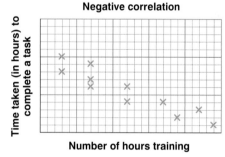

Figure 3.48

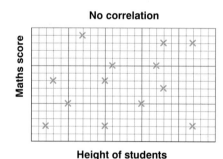

Figure 3.49

When a scatter chart shows positive or negative correlation, a *line of best fit* can be drawn. This is a line which follows the general trend of the points (sometimes called a *trend line*). There should be roughly as many points above a trend line as below it. The line can be used to make estimates about unknown values.

In the scatter graph, the line of best fit has been extended to cover a journey distance of 6 miles (see the dotted line). Using the extension, it seems reasonable to expect a 6 mile journey to take about 26 minutes.

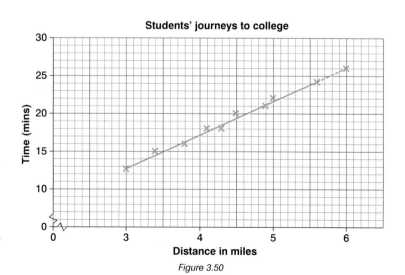

Figure 3.50

Now try these...

1 Draw a line of best fit for each scatter graph in the last exercise and state the type of correlation described by each scatter graph.

2 Give a common sense reason for the type of correlation in each graph in the previous exercise.

Take care when reading values from a line of best fit. Always ask yourself if the answer makes sense.

Example

A survey measures the population density of nine districts around a city. The distance from the city centre of each district is also recorded. The results are shown on the scatter graph below. A line of best fit has been drawn.

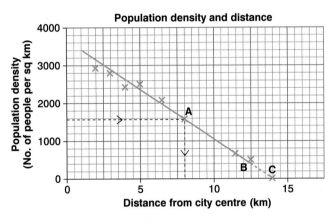

Figure 3.51

> **hint**
> 'Population density' can be calculated as the number of people per square km in a region.

a Briefly explain what the scatter graph indicates.

Districts *Charbridge* and *Stackton-Tressels* are not included in the survey. *Charbridge* has a population density of 1550 people per km². *Stackton-Tressels* is 14 km from the city centre.

b Use the line of best fit to estimate:
  i the distance of *Charbridge* from the city centre,
  ii the population density of *Stackton-Tressels* and comment on your answer.

Here's how...

a The graph shows **negative correlation**. This suggests the further away a district is from the city centre, the smaller its population density.

> **hint**
> Negative correlation means as one variable increases, the other decreases.

b i Go across from '1550' to point **A** on the line.
Then read down to '8 km'. **Answer:** 8 km

> **hint**
> Values found using the trend line are only estimates.

ii Extend the trend line from point **B** to point **C**. Point **C** indicates the population density is **zero**, i.e. there are no people living in *Stackton-Tressels*. Extending the line of best fit is not appropriate in this case.

> **hint**
> Always ask yourself if the answer makes sense.

Now try these...

1 The scatter graph shows the price of Thai rice
sold in bags of various weights.
a Explain briefly what the scatter graph
indicates.
b Use the line of best fit to estimate the pack
weight of a bag of Thai rice costing £4.

One of the bags is on special offer.
c Write down the price of this bag.

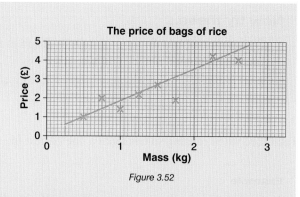

Figure 3.52

2 The scatter graph shows yearly attendance at
GCSE Maths re-sit classes and exam scores.
A line of best fit has been drawn.
a What does the scatter graph indicate?

One student missed a large number of classes
through illness but received extra tuition at
home.
b What was the percentage scored by this
student?

The pass mark for the course is 60%.
c What percentage of students passed the course?

Jackie attended for 95 hours but was unable to sit the exam.
d Use the line of best fit to estimate what Jackie's score would have been.

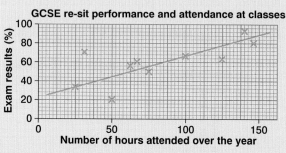

Figure 3.53

3 A farmer employs workers to gather a harvest. The scatter graph shows the connection (based on
previous experience) between the number of workers hired and the number of days taken to gather
the harvest.

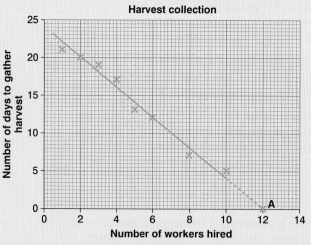

Figure 3.54

The harvest must be gathered in 10 days. The workers are each paid £50 per day.
a How many workers should the farmer hire in order to gather the harvest in 10 days?
b Use your answer to part **a** to find out how much in total the farmer should pay the workers.

The farmer decides to hire two extra workers. He must pay each worker a £20 bonus for every day
they finish ahead of schedule.
c How much money does the farmer save by hiring the extra workers?

The line of best fit has been extended to point **A**.
d Explain why point **A** has no sensible interpretation.

4 The scatter graph shows the amount of funding given to sports and the number of gold medals won at the 2000 Olympic Games by ten countries.

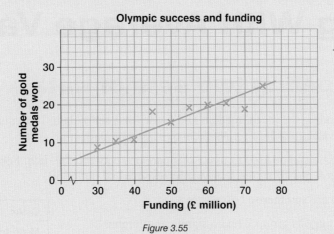

Figure 3.55

a One of the points on the scatter graph represents a country which won the least number of gold medals per £million funding. How much funding did that country receive?

b How much funding (to the nearest £million) is required for a country to win 30 gold medals?

Using the scatter graph, a sports official claims that even with no funding, a country can expect to win between four and five gold medals.

c Explain how the official used the graph to arrive at this conclusion. Comment on his claim.

5 The size of a police force in a county varies over 12 years. The scatter graphs show the size of the police force in a county and crime figures for each year.

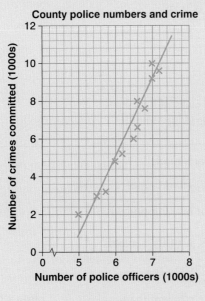

Figure 3.56

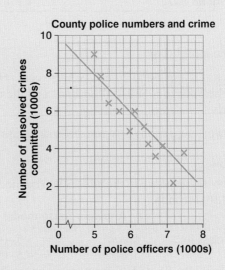

Figure 3.57

A detective is looking at one of the graphs and correctly states:

'*The more police officers there are, the more crimes that are solved.*'

Which graph is he looking at? Give a reason for your answer.

Working With Average Values

3.5.1 Comparing two sets of data

Chapter 3.3 deals with finding the mean, median and mode for a set of data. It is equally important to know how to interpret average values and how to use them to make comparisons.

The mean of two sets of data can be used to make comparisons between the data. For example, suppose two groups of students (group A and B) sit the same maths test marked out of 50.

The results of each group are summarised opposite:

The relatively low mean suggests group B had a significant number of low scoring students. On 'average', Group B found the test more difficult than group A.

Group	A	B
Mean score	37	21

Table 3.73: Summary of results

The mode and range also provide an easy way to compare two sets of results.

Example

The braking system of the new *Presto* car is being tested. A batch of 20 cars is selected. The time taken (in seconds) for each car to complete an emergency stop from a speed of 50 miles per hour is shown in the table.

Presto braking time (sec)	3.0–(3.5)	3.5–(4.0)	4.0–(4.5)
No. cars f	4	9	7

Table 3.74

a State the modal class.

The same test is applied to 20 new *Scherzo* cars. The modal class of braking times for the *Scherzo* is '4.5–5.0' seconds and the range of braking times is 4.2 seconds.

b Make two comparisons between the braking systems of each model.

Here's how...

a The highest frequency in the table is 9. The modal class of braking times is '3.5–(4.0)' seconds.

Answer: *Presto* modal class = '3.5–(4.0)' seconds

b Modal class for: *Presto*   = '3.5–(4.0)' seconds
        Scherzo   = '4.5–(5.0)' seconds

On average, the *Presto* braking system appears to be more efficient than that for the *Scherzo*.

Maximum range for: *Presto* times   ≈ 1.5 seconds
        Scherzo times   = 4.2 seconds

This suggests some of the *Scherzo* cars have relatively large braking times compared to *Presto* cars. Given the importance of braking times when driving, further testing of the *Scherzo* should be conducted.

hint

range = highest (possible) time − smallest (possible) time
= 4.5–3.0
= 1.5 seconds

Now try these...

1 12 students sit a physics test. The test is marked out of 50. The scores are:

$$23 \quad 34 \quad 25 \quad 42 \quad 30 \quad 41 \quad 19 \quad 35 \quad 47 \quad 22 \quad 29 \quad 37$$

   a Calculate:
   i the mean score   **ii** the range of scores.

A second group of students sit the same test. The second group has a mean score of 35 marks and a range of 14 marks.

   b Make two comparisons about the performance of the two groups.

2 The police measure the speed of eight cars passing a given point on the road where the speed limit is 50 mph:

$$\text{speed (mph)} \quad 55 \quad 45 \quad 51 \quad 42 \quad 59 \quad 62 \quad 54 \quad 60$$

   a Calculate:
   i the mean speed of the cars   **ii** the range of speeds recorded.

The police set up speed trap cameras by the side of the road. One week later, the mean speed of cars on the road was 49 mph. The range of speeds was 12 mph.

   b Have the cameras been effective in reducing driving speed? Discuss your answer.

3 pH values can be used to detect pollution in rivers. The pH value of pure water is 7. pH values below 7 indicate acidic water. pH values above 7 indicate that the water is alkaline.

A factory has recently started operating close to a river. A group of eight students are measuring the pH value of the river water. Each student takes five readings at different locations along the river. The results are:

pH level	4.5–	5.5–	6.5–	7.5–	8.5–(9.5)
No. readings f	20	10	5	0	5

Table 3.75

   a State the modal class.

One of the students used a faulty piece of equipment.

   b Which interval do you think shows this student's readings?

pH values in a similar location were taken one year before the factory was built. All the equipment used was in good working order. The readings had a modal class of 6.5–(7.5). The smallest reading as 6.3 and the largest was 7.6.

   c Is there evidence to suggest the factory is polluting the river? Carefully justify your answer.

4 A group of people gather on a hill to look for shooting stars. They record the time intervals (to the nearest 0.1 minutes) between observations of each shooting star, starting from the first sighting.

Waiting time: 2.3 1.5 2.2 3.4 0.3 1.1 1.6 3.0 0.2
              2.5 1.4 3.6 0.4 2.7 5.3 5.1 3.8 4.6

For example, the people waited 2.3 minutes to see the start of the next shooting star. The time for which each shooting star was visible can be ignored.

   a How many shooting stars were seen?
   b Calculate the mean waiting time between observations.
   c Calculate the range of waiting times between observations.

The next evening, the people return to the same spot. They observe shooting stars at a mean rate of two per minute. The range of waiting times between observations was 8.3 minutes.

   d Make two comparisons between the observations made on the two nights.

3.5.2 Finding information from bar-charts and histograms

The mean, median and mode can be found from a bar chart by reading off values from each bar.

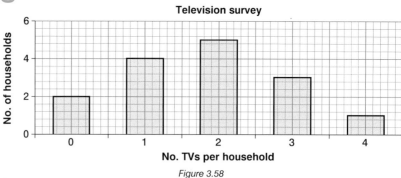

Figure 3.58

For example, suppose a survey on the number of television sets in 15 houses on a particular road is made. The results are shown using a bar-chart.

To find each average, it is often helpful to reconstruct the raw data on which the bar-chart is based.

The bar-chart shows that two households did not own a set, four households each owned one set and so on:

Raw data: No. sets per household: 0, 0, 1, 1, 1, 1, 2, **2**, 2, 2, 2, 3, 3, 3, 4

From the raw data, the middle value is 2 (highlighted) and the most common value is also 2. The **median** and the **mode** are both 2.

To find the mean, first calculate the total number of sets in the survey:

$$\text{Total no. of sets} = 0 \times 2 + 1 \times 4 + 2 \times 5 + 3 \times 3 + 4 \times 1$$
$$= 0 + 4 + 10 + 9 + 4$$
$$= 27$$

The mean number of sets per household $= \dfrac{\text{Total no. of sets}}{\text{Total no. households}} = \dfrac{27}{15}$

i.e. **mean** $= 1.8$

When finding the median, listing the values for small data sets is easy. For larger data sets you will need to keep a running total of frequencies so that you know when you have reached the 'middle' value.

Now try these...

1 Householders in a street are asked how many cats they own. The results are shown in the bar-chart.

a What is the modal number of cats per house?
b Calculate:
  i the number of houses in the street
  ii the total number of cats owned by the householders in the street
  iii the mean number of cats per house.
c Find the median number of cats.
d Which of the three averages do you think is the best representative for the number of cats per house?

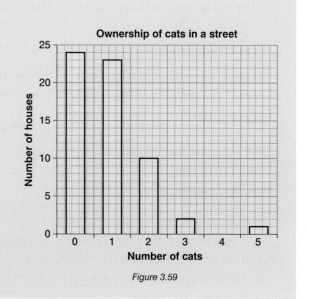

Figure 3.59

2 An awarding body carries out a survey to find the number of AS level subjects entered by A level students in the first year of their A level course. The bar-chart shows the survey results.

  a Estimate the percentage of students that study more than three AS level subjects.

  b Find

    i the mode

    ii the median

    iii the mean number of AS level subjects per student.

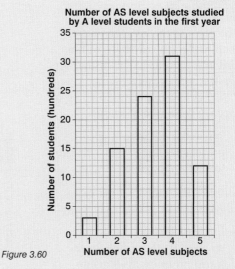

Figure 3.60

3 The bar-chart shows the number of cars sold by a salesman each day during one week.

  a Calculate the total number of cars he sold during the week.

  b Calculate the mean number of cars sold on weekdays (i.e. excluding the weekend).

  c Mark the following as true or false:

    i The salesman sold four times as many cars on Saturday than on Tuesday.

    ii The salesman sold 50% more cars on Saturday than on Friday.

    iii More than 50% of his sales were made during the weekend.

    iv The modal number of cars sold per day was 12.

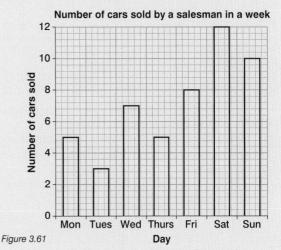

Figure 3.61

LEVEL 3

4 The histogram shows the time spent by a doctor in surgery consultations with his patients on one day.

  a Write down the modal class.

  b How many patients did the doctor see during the surgery?

  c What percentage of the patients had consultations lasting less than 4 minutes?

  d Use midpoints to estimate the mean consultation time per patient.
Why is your answer an estimate?

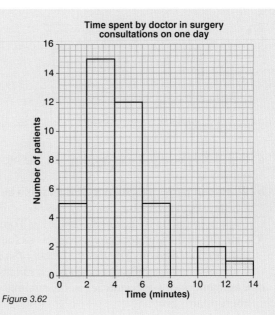

Figure 3.62

5 The mass of the luggage taken by each passenger travelling with an airline is recorded. The histogram illustrates the results for one year. Passengers can take up to 20 kg of luggage without any charge.

a What percentage of the passengers had to pay a charge for excess luggage?

The airline charges £2 per extra kilogram (over 20 kg) of luggage.

b Estimate the total amount received by the airline for the transport of excess luggage during the year.

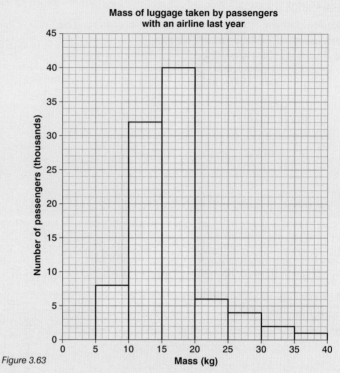

Figure 3.63

6 Tom and Jerry are two trainee accountants seeking promotion. Part of their job involves processing batches of invoices. The bar-charts show the number of mistakes made per batch for each trainee.

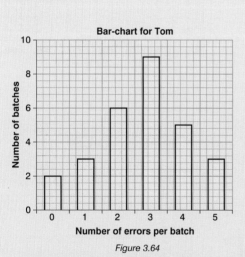

Figure 3.64

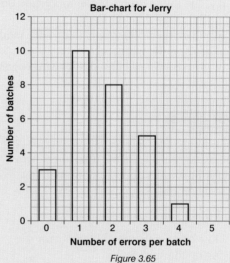

Figure 3.65

a Find **i** the mean **ii** the median **iii** the modal number of mistakes per batch made by each employee.

b When comparing reliability give a reason why:
i Jerry would prefer the mode was used **ii** Tom would prefer the median was used.

c Draw a component bar-chart combining Tom and Jerry's mistakes.

A summary of the component bar-chart is as follows:

  'The data is bimodal. The mean number of mistakes per batch is between 2 and 3 whilst the median is 1. The number of mistakes in a batch ranges from 3 to 14.'

d Comment on any inaccuracies in the summary. Support your answers with working where appropriate.

Portfolio Development

4.1 What is a portfolio?

The specifications for Application of Number Levels 2 and 3 are given on pages 180–83. Part A lists what you need to know how to do in order to produce the evidence in your portfolio. It also covers all the techniques that may be included in the external test. Your portfolio must contain at least one substantial piece of work. This is likely to be a project or investigation based on your studies at college or your life outside college including jobs, hobbies and interests. Some examples of possible assignments are given later in this chapter (Sections 4.6, 4.7 and 4.8). The file or folder that holds your project work is called your portfolio.

In the course of completing your portfolio for Application of Number you must:

- obtain and interpret information
- carry out calculations
- interpret answers and present findings.

You must also show how each of these activities relates to the aim of your project.

Everything you do must relate to the aim and purpose of your project.

Level 2

Look at Part B of the Application of Number specifications for Level 2 on page 181. This gives a detailed list of what you must do to complete your portfolio successfully.

Your portfolio must contain 100% of the Part B requirements. Check the details and ask your tutor if you need clarification.

It is a good idea to become familiar with the official numbering used in the specifications.
For example, in the notation **N2.1**
N stands for Application of **N**umber
2 stands for Level 2
.1 stands for 'Interpret information from **two** different sources, including material containing a graph'.

Your project must be a 'substantial' activity. This means that you must obtain and interpret information (N2.1), carry out calculations based on it (N2.2), interpret the results of these calculations and present your findings (N2.3), all within the same project. If you are not able to include all the categories of calculation in N2.2 in one project, then you must carry out further substantial activities, each meeting the requirements of N2.1 and N2.3 as well as N2.2.

For example, suppose in your chosen project, you are unable to cover 'handling statistics' (N2.2c). A separate page of mean, median and mode calculations taken out of context will ***not*** count as evidence for N2.2. Instead you would need to complete another project in which you obtained information then presented and interpreted your findings as well as carrying out the required statistical calculations. In view of this, it is important that you check with your teacher that your project is likely to produce the evidence you need before you start to work on it.

LEVEL 3

Look at Part B of the Application of Number specifications for Level 3 on page 183. As at Level 2, activities used to generate portfolio evidence must be substantial. This means that all the comments given above for Level 2 apply, with N2.1, N2.2 and N2.3 replaced by N3.1, N3.2 and N3.3 respectively.

In addition, activities used at Level 3 must be 'complex'. Complex activities are those which can be broken down into a series of tasks and where the techniques needed are more sophisticated. Assignments at level 3 will be more 'open-ended' giving rise to more than one possible approach and a variety of findings. A portfolio at this level will involve a lot of variables and a great deal of detailed work. It must also include evidence of planning, decisions about accuracy and justification of the methods used. The following general comments apply to both level 2 and level 3.

4.2 Organising your work

You will need to develop a system for collecting the information and evidence you need. Here is some general advice:

- *Start collecting information as soon as possible.*
  An A4 folder is ideal for collecting information you think might be useful. For example, if you find a newspaper article that might be relevant, cut it out or take a photocopy and pop it into your folder.

- *If you use information from a written source, keep a record of details* such as the author's name, the title of the book and when and where it was published. You will need to include these details when you write up your work. (See Section 4.4.4 about references.)

- *Keep a record of your planning.*
  Evidence of planning is advisable at Level 2 and essential at Level 3. Don't worry about getting the plan right first time. Jot down ideas as they occur to you, for example by 'brainstorming' with friends and colleagues. You can refine the plan later and organise it into a logical course of action. A well-structured plan with realistic target dates will help to reduce your stress levels!

- *Aim for quality rather than quantity.* When planning, think carefully about exactly what you are trying to do and how to do it in the most efficient way. Try to avoid unnecessary repetition.

- It may be difficult to store evidence for a particular type of work in a folder. For example, your project may involve making models or giving a presentation about your findings to a group. At the planning stage *think about what kind of evidence could be included in your portfolio*. You might, for instance, take photographs of your models or ask your tutor to make notes about your presentation.

- *Start a portfolio index (some awarding bodies suggest using a log book for this purpose).* This will show where each requirement listed in Part B of the specifications is met. Update the index after you complete each section of work. This will help you to see how well you are progressing and identify any gaps in what you have done. If a folder is not appropriate for storing evidence (for example, if you have made a model) your index should say where the evidence can be found.

- *Label all your work* with your name, your tutor's name, and the college address.

4.3 Obtaining information and data

4.3.1 Primary and secondary data

Data that you collect yourself is called **primary** data. This might consist of measurements you have taken in an experiment, observations you have made or results from a survey you have carried out. Collecting data yourself can be difficult and time-consuming. Consider other options before embarking on this course.

Secondary data is data that has been collected by someone else. Data found on the internet or in magazines or books is secondary data. Using secondary data is often preferable to using primary data because it is quicker to obtain and so releases extra time for working with the data.

4.3.2 Obtaining data

Planning

Think carefully about what it is you need to know, what sort of data you require, where you will find it and how you will make use of it. Choose the quickest and easiest way of getting the data you need.

 You do not need to use primary data.
You will probably save time if you use secondary data.

Using a Questionnaire

Asking people questions is a common way of collecting primary data. The questions are usually written down in a list to make sure everyone is asked questions in the same way. Such a list of questions is called a questionnaire.

Carrying out a survey using a questionnaire is a time-consuming process. Before you go ahead, consider whether there is an easier way of getting the information you need. If you still think that a questionnaire is the best method, follow the guidelines given below.

- *Planning the questionnaire*
  Decide what you want to find out and **brainstorm** the subject, making a list of the topics you wish to cover and writing down possible questions. Think about how you will analyse the answers.
  The quality of the data you get depends on how well you write your questions.

- *Guidelines for writing questions*
  Questions should be **short, simple and easy to understand**.
  Each question should have only **one possible meaning**.
  Questions should be **easy to answer**.
    A list of possible responses with boxes to tick or fill in makes answering easy and can also help you to analyse the results. If you are not sure whether you have catered for all possible answers include a box labelled 'Other' to cover any other possibilities.
    Give clear instructions about how you expect the question to be answered (eg 'Tick a box.')
  Avoid leading questions that make assumptions or introduce bias.
  Avoid personal or embarassing questions.
  Before making any copies of your questionnaire, ask a friend to help you by reading through the questions and telling you about any problems they have in understanding or answering them. Then make a final selection of questions from your list and re-write or re-order questions where necessary.

- *Overall layout*
  Your finished questionnaire should include:
    a **brief introduction** explaining the purpose of the questionnaire and how you will use the results
    a **polite request** for assistance including a **deadline** for return if necessary
    clear instructions on how to complete the questionnaire
    questions given in a **logical order**
    a brief **thank you**
    details of **how to return the questionnaire** if necessary.

- *Trial run*
  Try out your questionnaire on a few people before distributing it widely. This will allow you to see whether there are any problems and whether it is likely to give the data you require.

Populations and random samples

The population includes all the members of the group being considered. It could be a small group such as all the students on your course or a very large group such as all the teenagers in this country. In a **census** information is collected about every person living in the country. A full census of the UK population is carried out every ten years and results are available from internet websites. The most recent census was carried out in April 2001. Note that a population does not need to be a group of people. It could, for instance, be all the cars manufactured in the UK or all the houses in a particular town.

If you are collecting data yourself it is often impractical to include **all** members of the population concerned. If this is so, then information should be collected from a **random** sample that can be used to represent the population. A sample is random if every member of the population has an equal chance of being chosen. If a sample is not random then it is likely to be **biased** and any findings based on the sample may not be true for the population as a whole. For example, if the population consists of all the students in your college, you should not distribute questionnaires only to your friends or the students you know. They may have similar views to you and this would give a biased sample. To obtain a cross-section of opinions you could ask for a list of all students in the college and then use a random method to select a sample from the list. Many calculators have a key that will give random numbers to help you to do this. Ask your teacher for guidance.

In some situations it may not be possible for you to use a truly random sample. If this is the case, you should comment on this in your work, describing how your sample may be biased and in what ways this may have affected your results.

You will need to use enough data for statistical calculations to be meaningful, and for comparisons to be valid. Official guidance on the Level 3 specifications suggests that on at least one occasion you must work with 50 or more items. At Level 2, official guidance states that you must show that you can compare sets of data with a minimum of 20 items. Where possible work with even larger data sets, but remember that the most important thing is that your data must give relevant and meaningful results. It is acceptable to use slightly smaller data sets provided that this is for a worthwhile purpose.

Using an observation sheet

Before you start, think carefully about the type of information you will be collecting and design a suitable observation sheet. Give the sheet a title that briefly describes the data to be collected and make a note of any important details such as the date, time and place where the data is to be collected.

Some data may be collected as a list, but it often helps to use a table. (Section 3.2 gives more help with organising data using tally-tables.)

Using the Internet

Addresses of websites are called Universal Resource Locators (URLs). They are often given in newspapers, magazines and on television. If you know the address of the website that has the information you need, it is simply a matter of typing the address into the space provided at the top of the internet browser screen.

If you do not have a useful website address then you can use a search engine or on-line directory.

Clicking 'Search' on the internet browser screen gives you access to search engines. If you select a search engine and type in a key word or phrase, the search engine scans websites and gives a list of those that contain your word or phrase. A single search engine cannot keep track of everything, so it is often useful to try more than one search engine.

Take care when choosing which key word or phrase to use for the search. A general word such as 'football' will give you a list containing thousands of websites and finding the information you want will be like looking for a needle in a haystack. Target your search by using a more specific word or phrase such as 'football transfer fees'. Hints for searching can be found in the search engine's on-line Help facility.

On-line directories organise websites by category. Click on your main category of interest, for example Recreation and Sport, and you will be given a list of sub-categories such as Amusement and Theme Parks, Automotive, Aviation etc. You can progressively narrow down your search until you end up with a list of the most relevant websites.

Using a library

Most libraries have a wealth of resources including video and audio cassettes as well as books, journals and newspapers. Many also offer computer facilities with access to the internet and a collection of CD-ROMS. Find out what your college and local libraries have to offer and how you can use their cataloguing system to find the information you require.

4.4 Effective presentation techniques

It is important that all the work in your portfolio is presented clearly and logically.

4.4.1 Use of IT

You can write up your work by hand or use IT. Using word processing or desk-top publishing software has the advantage that the work can be used as evidence for the Key Skill in Information Technology as well as Application of Number. (The requirements for IT are listed at the end of Sections 4.6 and 4.7) However it is often difficult to create mathematical expressions on a computer. Do not devote too much time to sorting out problems of this kind. Consider using a mixture of both methods, word-processing the relatively straightforward sections and writing in any complex notation by hand.

Spreadsheets can be used to create impressive statistical charts and graphs. However it is vital that your Application of Number portfolio includes evidence to show that you understand the underlying calculations and that you could cope without the aid of technology. If you use a spreadsheet you must also give an explanation of the methods involved. For example, if you use a spreadsheet to draw a pie-chart you should explain how the angles of some of the sectors have been calculated. If you have drawn a line graph or bar-chart you could customise the default scales and labels and then describe what you have done and why.

If you use a spreadsheet for statistical calculations, use spreadsheet formulae to carry out the methods that you have learnt in Chapter 3, rather than the in-built statistical functions provided. A print-out of the formulae you have used will show that you understand the methods involved. Checking the calculations by hand would provide further evidence.

4.4.2 Use of tables

Tables provide a useful way of collecting data, setting out calculations and displaying results. When using a table, make sure that the columns and rows are clearly labelled and any units of measurement are clearly indicated. You should also give your table a title. If the table contains data collected by someone else (for example data from a newspaper or the internet) you must include the source. This is usually given below the table. (Section 4.4.4 gives advice about references.)

4.4.3 Graphs, charts and diagrams

It is a requirement of N2.3 and N3.3 that you use at least one graph, one chart and one diagram in your portfolio.

- *Graphs*
  include line graphs and scatter graphs. Graphs must be drawn using horizontal and vertical axes. A line graph may use more than one line. For example, a single graph can be used to show how the temperature at a number of resorts varies over the course of a day by using a separate line for each resort. A scatter graph also counts as a graph (with or without a line of best fit).

- *Charts*
  include pie-charts, bar-charts, histograms, frequency polygons, frequency charts or diagrams.

- *Diagrams*
  include scale drawings, plans, workshop drawings, circuit diagrams, 3D representations, flow charts, critical path or network diagrams and organisation charts.

Whatever illustrations you use, they must be drawn carefully, fully-labelled and give clear information.

4.4.4 General layout

The layout of your project or investigation should make it easy to follow. Include each of the following:

- *a cover sheet*
  Give the title of the piece of work, the date it was completed and the evidence it contains.
  Include spaces for the assessor's comments and signature when the work is marked.

- *any written instructions you were given*

- *an introduction*
  Describe the task in your own words, explaining the purpose of the activity.

- *an account of how you planned the work*
  Briefly describe the ideas you had and the approach you decided to use. At level 3 it is important to justify the methods you have used. If you discarded some of your ideas, say why.

- *an account of how you found the information* you needed
  It is important that you give reference details of your sources of information so that the assessor can check your work if necessary (see also bibliography below).

- *the work* you did, organised into *logical sections*
  Identify the level of accuracy in your work. For example, when measuring the dimensions of a room you might record the length and width to the nearest centimetre. If so, write this into your assignment and take account of this if you use the measurements in later calculations. At level 3 you will need to select appropriate levels of accuracy for yourself when tackling a task such as this.

- *checks* on calculations
  At each stage think about whether your results are making sense and include a note about this in your work. Check calculations by using estimations, approximations or reverse calculations (ToTT.8 gives help with this). In some cases you could use IT to check your work. For example, you could calculate the mean of a sample or draw a pie-chart by hand and then repeat the work using a spreadsheet as a check.
  Correct any errors you find.

- an *interpretation of each result*
  Always say what information is given by the result of a calculation, and give a brief description of what each of your charts and graphs show.

- a *conclusion*
  Summarise your findings and explain how these relate to the purpose of the activity.
  At level 3 you should discuss how your initial data and any assumptions you have made may have affected these findings. You could consider the size and structure of samples, the levels of accuracy you have used or the conditions in which you have done experiments. In the conclusion you could also make suggestions for further work that might be done to develop the project.

- a *bibliography*
  At the end of the work compile a list of the sources you have used. List author surnames in alphabetical order. You should also give the initial or first name of the author and the year of the publication in brackets. If the source is a book, give its title, the name of the publisher and where it was published.
  An appropriate layout is shown below:

Author	**Title**
Surname, Initial or First Name	Title of the Book
(Year of Publication)	Name of Publisher, Location

In the case of an article in a journal you should give the title of the article, the name of the publication, its volume number and its date as well as the publisher's name and the location:

Author
Surname, Initial or First Name
(Year of Publication)

Title
Title of the Article
Name of Publication, Volume Number, Date

When the piece of work is finished, number all the pages and sections of work so that it is easy to give the location of each piece of evidence you require for your portfolio. Ensure that the finished work is stored safely.

4.5 Portfolio checklist

When you are finally collecting all your evidence together in your portfolio you may find the following checklist helpful.

- **Portfolio Title Sheet**
  Give your name and the name of your teacher and school or college as well as the level of Application of Number at which the portfolio is to be assessed (eg Level 2 or Level 3).

- **List of Contents**
  If you have completed more than one piece of work it would be helpful to label these A, B, C etc, with pages of each piece of work numbered separately.

- **Index (or Log Book)**
  This should show where in your portfolio the evidence listed in Part B of the specifications can be found.

- **The completed pieces of work**

4.6 Level 2 sample assignment

Activities at Level 2 must be 'substantial'. This means that you must:

- obtain and interpret information
- carry out calculations
- interpret the results of these calculations and present your findings

all within the same project.

An example of an assignment and possible ways of approaching it are given below.

> Select two different locations. These could be villages, towns or areas of a city, with the two locations being similar or totally different in character. Investigate factors that you would consider before going to live in either of the places you have selected. Which of the two locations would you choose and why?

Choosing somewhere to live is something you will need to do at some point in your life and so this assignment has a real purpose. Even if you are presently living at home, it is likely that sooner or later you will want to rent or buy a place of your own. Perhaps your whole family is likely to be moving house soon. In this case your investigation could consider the advantages and disadvantages of two locations, bearing in mind the needs of all of your family members. Perhaps you are planning to go to university in the near future. If so, you could choose to compare student accommodation in places with universities. The same assignment tackled by different people can take many different forms.

The first step in tackling the investigation is to choose the two locations. You may choose locations that are similar (eg two small towns in the same county) or locations that are very different in character (eg a city centre and a village in the countryside).

You will then need to plan carefully what you want to find out, what information you need and how to get it. Some ideas about what you could include are listed below. You would not need to develop **all** of these ideas. Your aim should be to investigate the topic in a way that is relevant to you, whilst making the most of any opportunities that arise for producing suitable evidence for your portfolio.

Possible approaches

Cost of Housing *Aim: To compare the cost of renting or buying a property (eg a flat or a house).*	Evidence for your portfolio	Relevant sections in the book
Information about properties to rent or buy can be obtained from estate agents, local newspapers and internet websites. Decide what type of accommodation you are interested in and collect data about what is available in each of the two locations you have chosen. As well as the price, you could collect other types of data e.g. the number of rooms, what type of heating the property has, whether or not it has a garage etc. Each of your samples should contain **at least 20 properties**. Ideally you should use random samples, but this may not be practical. If so, discuss this when you write up your work explaining why you have used the samples you have selected. Comment on how this may have affected your results.	N2.1 Interpret information from **two** different sources	4.3 Using the Internet 3.2.2, 3.2.3 and 3.2.4 Organising data NB A practice data set is available from the Nelson Thornes website: www.nelsonthornes.com/gofigure 4.3 Random samples
Work out the mean, mode and median price and decide which of these is the best representative value. Calculate the range of prices. Use your results to make comparisons between prices in the two locations and explain how they would help you decide where to live. A spreadsheet can be used to calculate the averages and range. If you do use a spreadsheet, remember that you should also include some calculations done by hand to show that you understand the methods involved.	N2.2 Carry out calculations to do with: c handling statistics N2.3 Interpret the results of your calculations.	3.3.1, 3.3.3 and 3.5.1 Averages 3.3.2 Range
Draw charts to compare features of the properties in your samples e.g. you could draw histograms to compare prices, bar charts to compare the number of rooms, pie-charts to show the proportion of properties that have different types of heating, and so on. Again you could use a spreadsheet for some of this work, but you must support this with some hand-drawn charts. Explain how your results would help you decide where to live.	N2.3 Present your findings. Use at least **one** chart.	3.4.1, 3.4.2 and 3.5.2 Charts

House Price Inflation *Aim: To investigate how property prices change.*	Evidence for your portfolio	Relevant sections in the book
If you are considering buying a flat or house, you will be interested in how prices are likely to change in the future. An internet website gives information about how house and flat prices have changed in particular locations over the past few years. This information is given in graphs as well as in tables. Compare the way in which prices have changed recently in the two locations. Predict, with reasons, what you think is likely to happen in the next few years.	N2.1 Interpret information from **two** different sources, **including material containing a graph**.	4.3 Using the Internet The address of this and other useful websites are given on the Nelson Thornes website: www.nelsonthornes.com/gofigure
Draw line graphs to compare how house prices have changed in each location. Try to identify trends and predict what you think is likely to happen in the future.	N2.3 Present your findings. Use at least **one** graph.	3.4.3 Line Graphs

Floor Area *Aim: To compare the space in two properties*	Evidence for your portfolio	Relevant sections in the book
Estate agents' leaflets usually give the dimensions of each of the main rooms in a property that is for sale. Sometimes they include a detailed plan of each floor. Choose a property from each location with the same (or a similar) price. Ideally the properties should be similar in terms of their other features (e.g. either both have a garage or neither has a garage).	N2.1 Interpret information from **two** different sources.	2.3.3 Scale Drawings
Draw, to scale, a plan of each floor or each of the main rooms.	N2.3 Use at least **one** diagram.	2.3.3 Scale Drawings
Use area formulae (or count squares) to find the area of each room. Find the total floor area for each property. Use your results to make comparisons. The proportion of the floor space that is allocated to reception rooms, bedrooms and other parts of the property can be compared using ratios or percentages.	N2.2 Carry out calculations to do with: **a** amounts and sizes **b** scales and proportion **d** using formulae	2.3.1 Area 2.1.2 Ratios 2.1.3 Percentages
You could draw pie-charts to compare the proportion of space allocated to different parts of the properties. Use all of your findings to decide which property has the best layout and gives the best value for money (in terms of space).	N2.3 Interpret the results of your calculations and present your findings. Use at least **one** chart.	3.4.2 Pie-charts

Council Tax Aim: To compare community charges in the two locations.	Evidence for your portfolio	Relevant sections in the book
Find out about council tax rates in the two locations you have chosen. Properties are allocated to a number of 'bands' (A, B, C, ...etc). The council tax charged to people living in a property depends on which band the property is in. Find out what the council tax bands are by contacting the local councils or using the internet. Many councils maintain a website and the information is also available from other websites.	N2.1 Interpret information from **two** different sources.	4.3 Using the Internet Useful website addresses are given on the Nelson Thornes website: www.nelsonthornes.com/gofigure.
Draw a comparative bar-chart to illustrate and compare the council tax for each band.	N2.3 Present your findings. Use at least **one** chart.	3.1, 3.4.1 and 3.4.2 Charts

Crime Rates and Insurance Aim: To find out what the risks are and what effect this has on insurance rates.	Evidence for your portfolio	Relevant sections in the book
The Home Office website gives information about crime rates in different areas. This includes information about burglaries, car theft, assaults and other crimes. Other websites give information about crime and clear-up rates. Some information is given in the form of graphs and charts. Use this data to consider how likely you are to be the victim of a crime such as burglary or car theft in each of the locations you have chosen. Use the websites maintained by insurance companies to obtain quotes for house and/or car insurance in the locations you have chosen. Consider whether these quotes reflect your findings about crime rates. Describe how your results might affect your choice of where to live.	N2.1 Interpret information from **two** different sources, **including material containing a graph.**	4.3 Using the Internet Useful website addresses are given on the Nelson Thornes website: www.nelsonthornes.com/gofigure.

Job Opportunities Aim: To compare unemployment rates and employment opportunities	Evidence for your portfolio	Relevant sections in the book
Unemployment rates in different areas can be obtained from the National Statistics website. Find out what jobs are available near the two locations you have chosen. Sources of such information include job centres, advertisements in local newspapers and internet websites. Collect information about the jobs that would suit your ambitions and qualifications. If possible, use a sample for each location containing **at least 20 jobs**. Explain how you selected your samples.	N2.1 Interpret information from **two** different sources.	4.3 Using the Internet 3.2.2, 3.2.3 and 3.2.4 Organising data

Job Opportunities (cont.)	Evidence for your portfolio	Relevant sections in the book
Calculate the mean, mode and median wage and decide which of these is the best representative value. Calculate the range. Explain what relevance these calculations have to your choice of location. A spreadsheet can be used to calculate the averages and range. If you do use a spreadsheet, remember that you should also include some calculations done by hand to show that you understand the methods involved.	N2.2 Carry out calculations to do with: **c** handling statistics N2.3 Interpret the results of your calculations.	3.3.1, 3.3.3 and 3.5.1 Averages 3.3.2 Range
Draw charts to compare aspects of the jobs in your samples e.g. the availability of different types of jobs you are interested in, wages etc. Again you could use a spreadsheet for some of this work, but you must support this with some hand-drawn charts.	N2.3 Present your findings. Use at least **one** chart.	3.4.1 and 3.4.2 Charts

Cars, planes, buses, boats and trains. *Aim: To investigate the convenience and cost of transport systems.*	Evidence for your portfolio	Relevant sections in the book
If you use public transport, collect bus and train time-tables from the bus and rail companies that serve the locations you have chosen. Investigate prices and how regular the services are. Include journeys you are likely to make regularly (for example to attend college, go to work, go shopping etc) and also those you make occasionally, (for example when visiting relatives or going on holiday).	N2.1 Interpret information from **two** different sources.	2.1.6 Timetables
Work out how much you are likely to pay for public transport over the course of a year in each location. Explain how this would affect your decision on where to live.	N2.2 Carry out calculations to do with: **a** amounts and sizes N2.3 Interpret the results of your calculations.	
If you are likely to own a car, use maps to estimate the distance that you would have to travel to get to work, reach the nearest motorway, airport, channel port etc. Distinguish between journeys you are likely to make regularly and also those you would make occasionally.	N2.2 Carry out calculations to do with: **b** scales and proportion	2.3.3 Scale Drawings

Cars, planes, buses, boats and trains (cont.)	Evidence for your portfolio	Relevant sections in the book
If you are likely to travel by car, you could use formulae to work out what the difference in travelling times and costs would be from each location to the places you are likely to visit. You will need to find out the cost of petrol and estimate the speed at which you would travel and the distance your car would travel on each litre of petrol.	N2.2 Carry out calculations to do with: **a** amounts and sizes **b** scales and proportion **d** using formulae N2.3 Interpret the results of your calculations	2.2.1 Formulae
Leisure Activities *Aim: To find out what is available in the chosen locations.*	Evidence for your portfolio	Relevant sections in the book
Local councils often have websites that give information about facilities such as sports centres, cinemas and shopping centres. There are also websites maintained by all sorts of organisations involved in leisure activities (eg football clubs, rambling associations, clubs for various hobbies and interests). Find out what is available that would interest you in or near the locations you have chosen.	N2.1 Interpret information from **two** different sources.	4.3 Using the Internet
Education *Aim: To find out about opportunities and performance in education.*	Evidence for your portfolio	Relevant sections in the book
The Department for Education and Skills maintains a website that gives information and statistics related to education. For example, the data will allow you to compare the pass rates for GCSE, A level and vocational qualifications in different areas and within different schools and colleges in the same area. Information about universities can also be found on this website. There are other websites that also give data about education. Some of the information is in the form of tables, graphs and charts. Use this data to consider how good the educational opportunities are in the locations you have chosen.	N2.1 Interpret information from **two** different sources including material containing a **graph**.	Addresses of useful websites are given on the Nelson Thornes website: www.nelsonthornes.com/gofigure.
Medical Facilities *Aim: To compare hospitals, doctors and other medical facilities*	Evidence for your portfolio	Relevant sections in the book
The Department of Health has a website that gives data about hospitals and other medical facilities in different areas. This includes information such as the length of hospital waiting lists, ambulance response times to emergency calls and the doctor:patient ratio in general practice. Compare the data for your chosen locations and explain how this might affect your decision on where to live.	N2.1 Interpret information from **two** different sources.	Addresses of useful websites are given on the Nelson Thornes website: www.nelsonthornes.com/gofigure.

The **evidence** column in the table shows that there are several ways of meeting all the portfolio requirements for Application of Number within this assignment. It is vital at each stage to say why you are doing the work, explain what your results mean and how they will help you to reach decisions about where to live. You must also conclude the assignment with a summary, drawing together all the findings and giving an answer to the question about which location you would choose.

Information Technology

The table above identifies many opportunities in the assignment for using the internet and spreadsheets. A summary of the portfolio requirements for the Information Technology Key Skill at Level 2 is given below:

- **IT2.1**
  Search for and select information for **two** *different purposes.*
- **IT2.2**
  Explore and develop information, and derive new information, for **two** *different purposes.*
- **IT2.3**
  Present combined information for **two** *different purposes. Your work must include at least* **one** *example of text,* **one** *example of images and* **one** *example of numbers.*

(Consult the Information Technology Key Skill specifications for more detailed information about these requirements.)

Many of these requirements for IT could be met using the opportunities identified in this project. But remember that if you want your work to satisfy the Application of Number requirements as well as those for IT, you must use IT alongside work done by hand, rather than as a replacement for it. To meet the requirements for Application of Number you must prove to the assessor that you could carry out the methods **without** the aid of a computer.

Communication

The assignment described above could also make a major contribution to your portfolio for the Communication Key Skill Unit. Here is a summary of the requirements for Level 2:

- **C2.1a**
  Contribute to a group discussion about a straightforward subject.
- **C2.1b**
  Give a short talk about a straightforward topic, using an image.
- **C2.2**
  Read and summarise information from **two** *extended documents about a straightforward subject. One of these documents should include at least* **one** *image.*
- **C2.3**
  Write **two** *different types of documents about straightforward subjects. One piece of writing should be an extended document and include at least* **one** *image.*

(Consult the Communication Key Skill specifications for more detailed information about these requirements.)

Your assignment could be used as the extended document required for C2.3. If part of the project was discussed in a group at some stage and some of your results were given in a short talk you could also satisfy both requirements for C2.1. Information found on the Internet might help with C2.2.

Careful planning is needed if you wish to use one assignment to satisfy the majority of the requirements for all three key skills, but it is possible if you select an assignment such as this. However you may prefer to do a series of shorter assignments rather than one major assignment of this type. Some further suggestions for both long and short assignments are given in Section 4.8. Many of these are grouped according to subject area but this is not meant to be restrictive. You should aim to do an assignment that relates to your studies. However, if this is not possible, you should choose an assignment that relates to some other aspect of your life, for example a part-time job or your leisure interests.

4.7 Level 3 sample assignment

Activities at Level 3 must be 'substantial' and 'complex'. The project should be 'open-ended' i.e. have more than one possible approach and lead to a variety of findings. At this level your assignment will involve careful planning followed by a lot of detailed work. Your portfolio must contain evidence of decisions you make about levels of accuracy. You must also justify the methods you use and show how they meet the aim of the task.

An example of an open-ended assignment and possible ways of approaching it are given below. It involves something you will need to do at some point in your life (even if not right now) and so has a real purpose.

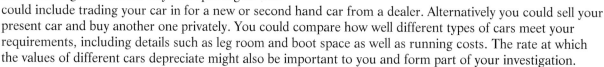

Investigate factors that you think you should take into account before buying a car.
Which car (if any) would you choose and why?

The first step in tackling any investigation is to plan carefully what you want to find out, what information you will need and how to get it.

Your approach to this particular assignment will depend on your personal circumstances. If you have not yet learned to drive, you could start by investigating the number of lessons you are likely to need to pass the driving test and the costs involved. You might then look in detail at the types of cars available, how much they cost and consider alternative ways of paying for one.

If you have already learned to drive and own a car you could investigate different ways of replacing it. This could include trading your car in for a new or second hand car from a dealer. Alternatively you could sell your present car and buy another one privately. You could compare how well different types of cars meet your requirements, including details such as leg room and boot space as well as running costs. The rate at which the values of different cars depreciate might also be important to you and form part of your investigation.

You may think that you cannot afford to buy a car at present because your income is very low or irregular. If so, you could include a study of the income you are likely to receive in your future career. You could then look at the factors that would influence the sort of car you would then wish to buy. Alternatively you could compare the cost of buying and running a car with the cost and practicalities of using public transport. This may well confirm your belief that a car would cost too much for you at present.

There may well be other reasons why you would not choose to own a car. Perhaps you are concerned about the environment and the levels of pollution caused by cars. A detailed study of this aspect could form the basis of your project.

The same assignment tackled by different people can take many different forms. Some ideas about what *could* be included in this particular investigation are listed below. Your main aim should be to investigate the topic in a way that is relevant to you whilst making the most of any opportunities that arise for producing suitable evidence for your portfolio.

Possible Approaches

Passing the Driving Test *Aim: To estimate the cost of learning to drive and how long it would take.*	Evidence for your portfolio	Relevant sections in the book
Select a sample of driving schools from local newspapers, a telephone directory or the Internet. Contact them and ask for information about the price of lessons, the cost of taking the test and how many lessons and attempts at the test learners usually need. Alternatively carry out a survey amongst drivers who have recently passed their test. Ask your local test driving centre if they can supply you with information about pass rates.	N3.1 Plan and interpret information from **two** different types of sources.	4.3 Selecting a random sample 3.2 Organising data
Use the data collected to work out the 'average' price of driving lessons and the 'average' number of lessons and test attempts that learners need. Estimate how much it would cost to learn to drive and how long it would take.	N3.2 Multi-stage calculations: c handling statistics	3.3.1, 3.3.3 and 3.5.1 Averages
Use charts to illustrate the data. For example you could use a pie-chart or bar-chart to show the number of attempts required to pass the test and a frequency polygon to illustrate the total amount spent by a sample of drivers.	N3.3 Use at least **one** chart.	3.1, 3.4.1, 3.4.2 and 3.5.2 Charts
Draw line graphs to compare the costs of lessons from different driving schools. Such graphs can be produced using a spreadsheet, but draw at least one by hand to show that you understand the techniques involved. Use the graphs to decide which driving school you would use, explaining your reasoning.	N3.3 Use at least **one** graph.	3.4.3 Line graphs
Try to write formulae, in words or using letters, showing how the cost depends on the number of lessons taken. (Ask for help with this if necessary.) Rearrange your formulae to find the number of lessons you could buy from different driving schools for a fixed amount of money.	N3.2 Multi-stage calculations: d rearranging and using formulae	2.2.1 and 2.2.6 Formulae
Cost of a Car *Aim: To find out how much you would need to pay for a car you like or to compare prices for different cars.*	**Evidence for your portfolio**	**Relevant sections in the book**
Car prices for both new and used cars can be obtained from car dealers, local newspapers, publications from the car trade and internet websites. The data is very complex, taking a variety of forms. Some car advertisements may give only the model and price whilst others may also give the year of production, number of doors, colour, number of miles travelled and so on.	N3.1 Plan, and interpret information from **two** different types of sources including a **large data set**.	4.3 Using the Internet NB: Datasets for *practice* are available from the Nelson Thornes website (but you must find the data that you use in your portfolio yourself).

Cost of a Car (cont.)	Evidence for your portfolio	Relevant sections in the book
It is advisable to study prices and some of the other details for just a few different models rather than trying to cover everything. If the models you choose are relatively common there are likely to be a large number on sale. This will give you the opportunity of working with a **large data set** (i.e. at least 50 items). You may have a particular model of car in mind and wish to compare prices for different sizes of engine or compare the price of a diesel version with a petrol version. Alternatively you may wish to compare prices of similar types of cars or the price of a large car with that of a small car. Ideally you should use a random sample of the cars that you are interested in, but this may not be practical. If so discuss this when you write up your work and explain why you have used the sample you have selected. Consider any bias that may have been introduced and comment on how this may have affected your results.		3.2 Organising data 4.3 Random samples
For the model(s) you have chosen work out the mean, mode and median price (to an appropriate degree of accuracy) and decide which of these gives the best representative price. Calculate the range. Use your results to make comparisons. Some dealers may offer a discount on a particular model. You will need to allow for this so that you can compare like with like. A spreadsheet can be used to calculate averages and measures of spread. If you do use a spreadsheet, remember that you should also include some calculations done by hand to show that you understand the methods involved.	N3.2 Multi-stage calculations: c handling statistics	3.3.1, 3.3.3 and 3.5.1 Averages 3.3.2 Range
Draw charts to compare features of the cars that are for sale. Again you could use a spreadsheet for some of this work, but you must support this with some hand-drawn charts.	N3.3 Use at least **one** chart.	3.4.1, 3.4.2 and 3.5.2 Charts
Depreciation *Aim: To find out how quickly a car loses its value.*	Evidence for your portfolio	Relevant sections in the book
If you are likely to sell your car at a later date it will be important to estimate how quickly your car will depreciate in value. This can be done by plotting a scatter graph of the price asked for second-hand cars against age. Draw the line of best fit and use it to find the rate at which cars lose value and make predictions for your car.	N3.3 Use at least **one** graph	3.4.4 Scatter graphs

Depreciation (cont.)	Evidence for your portfolio	Relevant sections in the book
A spreadsheet can be used to check your scatter graph. The spreadsheet may give the equation of the line of best fit (sometimes called the 'trendline'). This gives a formula relating the value of a car to its age. The formula could be used to predict the value of a car at various ages or can be rearranged to give the age of the car in terms of its value. You can then apply these results to your car.	N3.2 Multi-stage calculations: **d** rearranging and using formulae	2.2.1 and 2.2.6 Formulae

Finding the money *Aim: To find out how long it would take you to save enough money to buy a car, what it would cost to take out a loan, or to compare different ways of paying for a car.*	Evidence for your portfolio	Relevant sections in the book
Collect leaflets from banks and building societies or visit their websites. Take account of any special deals or discounts so that you can compare like with like.	N3.1 Plan, and interpret information from **two** different types of sources.	4.3 Using the Internet
Estimate how much you would need to save each month to have enough to buy a car within a year or two years (bearing in mind that the cost of the car might change in that time). This work may include percentage calculations or the use of formulae. Investigate the cost of taking out a loan and work out how much extra you would pay in interest. Compare the cost of a loan from a bank or building society with the deals offered by car dealers. Consider using a spreadsheet to perform the calculations involved, especially when they are repetitive. A spreadsheet can be used to investigate quickly the effect of altering the amount you save each month or the rate of interest. It also provides an opportunity of working with formulae. It is useful to print out the formulae you use as well as the numerical results. Remember that you should check some of the calculations by hand to show that you can also work things out for yourself without the aid of the spreadsheet.	N3.2 Multi-stage calculations: **a** amounts and sizes **d** rearranging and using formulae	2.1.3 Percentage calculations 2.2.1 Formulae

Running Costs *Aim: To estimate the total cost of running a car, to compare running costs for different cars or to compare the cost of running a car with that of public transport.*	Evidence for your portfolio	Relevant sections in the book
Find out likely maintenance costs for your preferred car(s). Include occasional costs such as the price of replacement tyres, brakes and an exhaust system as well as the costs of regular servicing. Add the cost of road tax and an MOT if the car is over 3 years old.	N3.1 Plan, and interpret information from **two** different types of sources.	4.3 Using the Internet

Running Costs (cont.)	Evidence for your portfolio	Relevant sections in the book
Information about prices can be found from other drivers or the internet as well as local garages, adverts in newspapers or magazines and catalogues of replacement parts. Another running cost is car insurance. Use the internet or an insurance broker to find out how much it would cost to insure your preferred car(s) for different levels of cover. Include details of no-claims bonus and other discounts.		
Work out how far you are likely to travel in a year and the cost of the petrol (or diesel) you will use. Car manufacturers' brochures and magazines give the number of miles to the gallon travelled by different makes of car on various types of journey. You will also need the current cost of fuel and the conversion factor for gallons and litres. Include the cost of regular and other journeys you make, for example, when going on holiday or visiting friends and relatives. Write a list of all the costs you have found. Add these up to estimate the total annual cost of running a car. Costing the same journeys on public transport would allow you to compare the annual costs. You could extend this work to compare estimates of the time it would take to make these journeys by car with the time taken by public transport.	N3.2 Multi-stage calculations: **a** amounts and sizes **b** scales and proportion	2.1.2 Proportion 2.1.6 Converting units 2.2.2 Compound measures
Draw charts to illustrate costs. For example you could draw a pie-chart to compare the different costs involved in running a car. Or you could draw a comparative bar-chart to compare the annual cost of using a car for the different types of journeys you make with the cost of using public transport.	N3.3 Use at least **one** chart	3.1, 3.4.1 and 3.4.2 Charts
A spreadsheet can be used to model a car account over the course of a year. Start by estimating the average amount you can afford to spend on the car per month and set up the spreadsheet so that this amount is added to the account at the beginning of each month. Subtract an estimate of the amount needed to pay for petrol each month. Make other deductions for oil, MOT, road tax, insurance and any other costs you think should be included. Use the spreadsheet to find the balance of the account at the end of each month. By using a spreadsheet you can easily investigate the effect of varying the monthly payments or making a deduction for a costly repair.	N3.2 Multi-stage calculations: **a** amounts and sizes **d** rearranging and using formulae	

Running Costs (cont.)	Evidence for your portfolio	Relevant sections in the book
Line graphs can be drawn to show how the balance varies during the year. Remember that you must check some of the calculations and draw at least one graph by hand to show that you could cope without the spreadsheet's help.	N3.3 Use at least **one** graph	3.4.3 Line graphs

Salaries *Aim: To find out how much you are likely to earn in future.*	Evidence for your portfolio	Relevant sections in the book
Look at job adverts for a career of your choice and collect data about the salaries paid. This may provide an opportunity for working with a **large data set (i.e. at least 50 items)**.	N3.1 Plan, and interpret information from **two** different types of sources.	3.2.2, 3.2.3 and 3.2.4 Organising data
Calculate the mean, mode and median salary and decide which of these is the best representative value. You may want to distinguish between starting salary and final salary as well as between gross and net salaries. Calculate the range. Explain what relevance these calculations have to your choice of car.	N3.2 Multi-stage calculations: c handling statistics	3.3.1, 3.3.3 and 3.5.1 Averages 3.3.2 Range

Performance Data *Aim: To compare, for example, how quickly different cars accelerate and how long they take to stop.*	Evidence for your portfolio	Relevant sections in the book
Performance data for different models of cars can be obtained from car dealers, manufacturers' websites or car magazines. Some of this data is given in tabular and graphical form. You could compare and contrast such information for two or more different models. An investigation of the way in which the efficiency of your car (in miles per gallon) varies with speed might allow the optimum speed to be found. The most efficient speed for different models could be compared.	N3.1 Plan, and interpret information from **two** different types of sources.	4.3 Using the Internet

Space in Cars *Aim: To compare the space available for passengers and luggage in different cars.*	Evidence for your portfolio	Relevant sections in the book
If possible take measurements from actual cars. If this is not possible, try to find out the information you need from a car dealer. Manufacturers' brochures and websites sometimes give scale drawings of the plan and elevations of a car as well as the main dimensions.	N3.1 Plan, and interpret information from **two** different types of sources.	4.3 Using the Internet

Space in Cars (cont.)	Evidence for your portfolio	Relevant sections in the book
Compare dimensions such as seat widths, legroom and seat to roof height for different models of cars. Take your own measurements to see how much room you would have to spare when sitting in the driver's seat. Take measurements from family and friends to see how well they would fit into the passenger seats.		2.3.3 Scale diagrams
Construct scale drawings to show the shape and dimensions of passenger compartments and boots.	N3.3 Use at least **one** diagram.	2.3.3 Scale drawings
Estimate the volume of the passenger compartments and boots. This can be done by dividing the space into shapes such as cuboids for which the volume can be easily found. Compare the volumes for different models of cars. Consider whether the boot space will be adequate for your needs.	N3.2 Multi-stage calculations: **a** amounts and sizes **b** scales and proportion **d** rearranging and using formulae	2.3.2 Volume

Window Areas *Aim: To compare visibility from different models of car.*	Evidence for your portfolio	Relevant sections in the book
If possible measure the dimensions of the cars' windows or find the dimensions from manufacturers' brochures.	N3.1 Plan, and interpret information from **two** different types of sources.	
Draw a sketch of each window.	N3.3 Use at least **one** diagram.	
Estimate the area of each window. This can be done by approximating the shape of the window using a rectangle, triangle or trapezium. Estimate also the area of the metal uprights that lie between the windows. Write the area of the windows as a percentage of the total area 'all-round' area. Compare percentages for different models of car and explain how this will affect the driver.	N3.2 Multi-stage calculations: **a** amounts and sizes **b** scales and proportion **d** rearranging and using formulae	2.3.1 Area 2.1.3 Percentages

Keeping the car in a garage *Aim: To find out how well a car will fit into your garage or to estimate the cost of building a garage*	Evidence for your portfolio	Relevant sections in the book
If you have already got a garage it is obviously important to check that any car you buy will fit into it. Measure the dimensions of your garage. Find out the main dimensions of a car you like (by measuring or from a dealer or manufacturer's brochure).	N3.1 Plan, and interpret information from **two** different types of sources.	
Draw a sketch or scale diagram showing the car inside the garage. Work out how much space there will be to spare, how far you will be able to open the doors etc.	N3.3 Use at least **one** diagram.	2.3.3 Scale diagrams.
If you have not got a garage you may wish to design one, especially if the cars you are considering are very expensive. Find out the main dimensions of a car you like (by measuring or from a dealer or manufacturer's brochure).	N3.1 Plan, and interpret information from **two** different types of sources.	
Work out the dimensions for your garage allowing space around the car for opening doors etc. Draw a sketch or scale drawing of the garage.	N3.3 Use at least **one** diagram.	2.3.3 Scale diagrams.
Calculate the total cost of building the garage. This may be a complex activity that requires you to calculate areas and volumes as well as doing money calculations for the cost of materials and labour. The work could be extended to include the purchase and storage of tools and a work bench if you wish to maintain the car yourself. You could consult DIY catalogues for this information.	N3.2 Multi-stage calculations: **a** amounts and sizes **b** scales and proportion **d** rearranging and using formulae	2.3.1 Perimeter & area 2.3.2 Volume 2.1.2 Ratio & proportion 2.1.3 Percentages (VAT)

Danger on the Roads *Aim: To find out what the risks are on the road.*	Evidence for your portfolio	Relevant sections in the book
The Department of Transport, Local Government and the Regions has a website that gives information about transport accident statistics. Some of this information is in the form of graphs and charts. Use this data to consider how safe you will be on the roads. For example, the data will allow you to compare accident rates for males and females and how safe a cyclist is in comparison with the driver of a car.	N3.1 Plan, and interpret information from **two** different types of sources.	Addresses of useful websites are given on the Nelson Thornes website: www.nelsonthornes.com/gofigure.

Pollution caused by cars *Aim: To find out what damage is done to the atmosphere.*	Evidence for your portfolio	Relevant sections in the book
Data relating to pollution levels and the contribution made by cars can be found on websites maintained by the Government and Environmental Agencies. Some of the information is given in tables, graphs and charts. Use this data to consider how much damage to the atmosphere is caused by the use of cars and what the consequences of this damage is likely to be. Make predictions about the future, stating any assumptions you make.	N3.1 Plan, and interpret information from **two** different types of sources.	Addresses of useful websites are given on the Nelson Thornes website: www.nelsonthornes.com/gofigure.

The **evidence** column in the table shows that there are several ways of meeting all the portfolio requirements for Application of Number within this assignment. Your project does not need to develop **all** the ideas listed, but does need to be a major piece of work with many inter-related aspects. It is vital at each stage to say why you are doing the work, explain what your results mean and how they will help you to reach decisions about buying a car. You must use an appropriate level of accuracy and conclude the assignment with a summary, drawing together all the findings and giving an answer to the question about which car (if any) you would choose to buy.

Information Technology

The table above identifies many opportunities in the assignment for using spreadsheets and the Internet. A summary of the portfolio requirements for the Information Technology Key Skill at Level 3 is given below:

- **IT3.1**
  Plan and use different sources to search for, and select, information required for **two** *different purposes.*

- **IT3.2**
  Explore, develop and exchange information, and derive new information, to meet **two** *different purposes.*

- **IT3.3**
  Present information from different sources for **two** *different purposes and audiences. Your work must include at least* **one** *example of text,* **one** *example of images and* **one** *example of numbers.*

(Consult the Information Technology Key Skill specifications for more detailed information about these requirements.)

The majority of these requirements for IT could be met using the opportunities identified in this one project. But remember that if you wish your work to satisfy the Application of Number requirements as well as those for IT, you must use IT alongside work done by hand, rather than as a replacement for it. Remember that to meet the requirements for Application of Number you must prove to the assessor that you could carry out the methods without the aid of a computer.

Communication

The assignment described above could also make a major contribution to your portfolio for the Communication Key Skill Unit. Here is a summary of the requirements for Level 3:

- **C3.1a**
  Contribute to a group discussion about a complex subject.

- **C3.1b**
  Make a presentation about a complex subject, using at least **one** *image to illustrate complex points.*

- **C3.2**
  Read and synthesise information from **two** *extended documents about a complex subject. One of these documents should include at least* **one** *image.*

- **C3.3**
Write **two** *different types of documents about complex subjects. One piece of writing should be an extended document and include at least* **one** *image.*

(Consult the Communication Key Skill specifications for more detailed information about these requirements.)

The assignment could be used as the extended document required for C3.3. If the project was discussed in a group at some stage and the results were given in a presentation you could also satisfy both requirements for C3.1. An article in a car magazine or information found on the Internet might help with C3.2.

Careful planning is needed if you wish to use one assignment to satisfy the majority of the requirements for all three key skills, but it is possible if you select an open-ended assignment such as this. However you may prefer to do more than one assignment, each being a substantial and complex activity covering some of the requirements. Further suggestions for assignments are given in the following section. Many of these are grouped according to subject area but this is not meant to be restrictive. You should aim to do an assignment that relates to your studies. However, if this is not possible, you should choose an assignment that relates to some other aspect of your life, for example a part-time job or your leisure interests.

4.8 More assignment ideas

General

Plan the refurbishment of a room.

Compare salaries for different jobs.

Plan a camping (or other) holiday.

Plan a garden.

Investigate which is the best mode of transport.

Plan a sponsored walk.

Compare mobile phones.

Art and Design

Plan an exhibition or fashion show

Design and cost an outfit or a piece of furniture or equipment.

Plan a building to be used as a youth club.

Which artist is the most popular?
What makes the artist's work popular?

Who is the most well-known fashion designer and why?

Catering

Plan a party or wedding reception.

Is vegetarianism becoming more popular? Investigate.

Plan a snack bar for students in your college.

Compare the eating habits of the 'average' student with those recommended for healthy eating.

Leisure and Tourism

Design a swimming pool or sports centre.

Compare the popularity of different sports (or hobbies).

Plan a holiday camp.

Plan a theme park.

Organise a concert or play.

Compare two tourist attractions.

Compare the efficiency and services provided by two bus/coach companies.

Engineering

Design a workshop.

Draw a plan and elevations of a piece of equipment.

Design and compare the advantages and disadvantages of two container shapes.

Compare the training requirements and salaries for different engineering jobs.

Health and Social Care

How healthy is vegetarianism?

What proportion of people are overweight?

Plan a health centre.

Compare life expectancy and 'quality of life' in different countries.

Compare the health of men and women in this country.

Compare health or caring facilities in different areas.

Economics/Business Studies

Do men earn more than women?

Businesses either grow or shrink – they don't stay the same size. Investigate how true this is.

Do we pay too much tax?

Should petrol duty be reduced?

What are the financial implications of being a student?

Devise a business plan for a business of your choice.

Computing

Who uses the Internet most and why?

Buying a home computer – what factors should be considered?

Compare the popularity of computer games with other pastimes.

How has the use of computers changed the type of work we do?

To what extent is industry computerised?

Plan and cost a work station.

Humanities and Social Sciences

Does telepathy exist?

Investigate the effects that charging university fees has had on the student population.

Compare the standard of living in the UK with that in a third world country.

Quality of life is better in the country than in the town. Investigate.

Compare population densities in different countries.

Compare the amount given to charities to the amount spent on gambling.

Science

Soaking seeds in water before planting helps them to germinate. Investigate.

How far away from the parent plant do seedlings grow?

Compare the advantages and disadvantages of being a large animal, rather than a small animal.

What evidence is there for global warming?

Once you have decided on an assignment check with your Application of Number tutor to make sure the project is suitable before proceeding.

Having chosen an assignment, small group or class discussion can help to generate ideas.

The External Test

5.1 What will the test be like?

The test papers are the same whichever awarding body (AQA, Edexcel or OCR) your school or college is registered with. Exemplar tests and past papers can be downloaded from the websites maintained by QCA (Qualifications and Curriculum Authority) and the awarding bodies.

Level 2

A level 2 test lasts for $1\frac{1}{4}$ hours and contains forty multiple-choice questions, each worth one mark. For each question you are given four possible answers, A, B, C and D. You are required to choose the correct option and mark the letter on a separate answer sheet. Calculators are **not** allowed in the level 2 test.

Level 3

A level 3 test lasts for $1\frac{1}{2}$ hours. All questions are based on real-life contexts. Section A consists of short questions; Section B is an extended question based around one scenario and broken down into a series of sub-questions. Answers are written in a separate answer booklet and on separate sheets of graph paper or squared paper when necessary. The test is marked out of 50 with approximately half of the marks in Section B. Scientific calculators are allowed in the level 3 test.

5.2 Revising for the test

- Ensure that you **know what you will be tested on**. A list of what you should know is given in Section A of the Application of Number Specifications (see pages 180–183).

- Find out the date of the test and **plan a revision timetable**. You could start by revising one topic at a time, but allow plenty of time at the end for trying mixed questions and past papers. Allow time for breaks, relaxation and your other interests.

- **Set yourself achievable targets**.
  Once you start, tick off what you achieve – this will help you feel better!

- Write a list of the formulae you need to know.
  Learn these formulae and **test yourself** (for example by writing out the formulae from memory).

- **Revise actively** by working through questions (including past papers).

5.3 Tackling the Level 2 test

- **Don't panic!**
  Easier said than done! – but try to stay calm. It will help you think more clearly.

- **Read each question carefully**.
  It may help if you underline important information.

- **If you do not know how to do a question, consider which of the given answers looks most likely**.
  Ask yourself if any of the given answers (A, B, C or D) are obviously wrong. If you have to guess an answer, try to guess sensibly!

- **If you decide to leave a question until later, mark it clearly** on the question paper so you can find it easily. Take care to give your answer to the next question in the correct place on the answer sheet. It is easy to make a mistake and get 'out of step'.

- Use any time you have left at the end of the test, to **check your answers**. If you decide to change an answer, ensure that you do this correctly, using the instructions on the answer sheet.

5.4 Tackling the Level 3 test

- **Don't panic!**
  Easier said than done! – but try to stay calm. It will help you think more clearly.

- **Read each question carefully**.
  It may help if you underline important information.

- **Take care with the presentation of your work**.
  Examiners must be able to read and follow it. Underline your answers to make them obvious.

- **Do not round values in the middle of a calculation**.
  This leads to inaccuracy and you will lose marks.

- **Give answers using appropriate units**.
  Marks are deducted for incorrect or no units.

- **Round each final answer sensibly**.
  In money calculations this is usually to the nearest pence. Give an answer as £1.70 rather than £1.7.
  If a question is about a number of items, round your answer to a whole number.
  (See ToTT.8 for more help with accuracy and rounding.)

- **Give charts and graphs a title**. (You will throw a mark away if you forget.)

- **Label axes clearly**. Include units if there are any.

- **Choose and use scales carefully**.
  Make sure your graph, chart or diagram will fit on the page.

- **Consider the marks allocated to each question**.
  If you get stuck, don't spend too long on a question that is only worth a few marks. Leave a space and return to it later if you have time. If you decide to leave a question until later, mark it clearly on the question paper so you can find it easily.

- Ensure that you **leave enough time for the extended question** in Section B. Remember that it carries about half of the total marks.

- Try to leave enough time at the end of the test to **check your answers**.

- If you redo a question, **make sure that the examiner can tell which solution you want to be marked** – cross out the other solution.

5.5 Sample test questions for level 2

The questions are in groups, each group set in a real context. Each question is worth one mark. You are required to select the correct answer from four possible results.

hint

Here each group of questions is clearly identified and labelled with the context. In a level 2 test the groups of questions are not identified so clearly.

Planning a Drive

Beth is planning a new concrete drive. The sketch shows a plan of the drive.

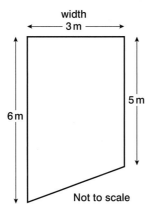

Figure 5.1

1 Beth draws the plan to a scale of 1:50. What is the width of the drive on her plan?

   A 1.5 cm   **B** 6 cm   **C** 12 cm   **D** 15 cm

2 The concrete is to be 50 cm deep. What volume of concrete is required?

   A 8.25 m^3 **B** 9 m^3 **C** 16.5 m^3 **D** 82.5 m^3

Here's how...

1 Actual width = 3 m = 300 cm.

The scale 1:50 means that any distance on the plan can be found by dividing the actual distance by 50.

$$\text{Width on plan} = \frac{300}{50} = 6 \text{ cm}$$

Answer = B

hint

Note the depth is in centimetres. This will have to be converted to metres.

hint

See Section 2.3.3 working with scales.

2 Area of concrete
= area of rectangle + area of triangle

$$= 3 \times 5 + \frac{3 \times 1}{2}$$

$$= 15 \quad + 1.5$$

$$= 16.5 \text{ m}^2$$

Volume of concrete
= area × depth
= 16.5 × 0.5 (as 50 cm = 0.5 m)
= 8.25 m^3

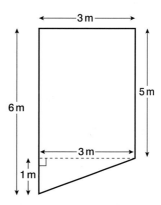

Figure 5.2

Answer = A

hint

See Section 2.3.1 for areas.

hint

See Section 2.3.2 for volumes.

Travelling Salesman

Stan is a travelling salesman for a company based in Bristol. When he visits customers he uses this chart to calculate the distances he travels.

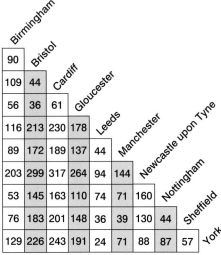

Mileage chart

Figure 5.3

Sometimes information is given that is needed in more than one question. This mileage chart is needed in questions 1 and 2 below.

Stan estimates the time it will take to make a journey using the formula:

$$\text{Time} = \frac{\text{Distance}}{\text{Speed}}.$$ He uses an estimated speed of 50 miles per hour.

See Section 2.2.2 for speed.

1 Which of these is nearest to Stan's estimate for the time it will take to travel from Bristol to Manchester?

  A $2\frac{1}{2}$ hours   **B** 3 hours   **C** $3\frac{1}{2}$ hours   **D** 4 hours

2 One day Stan goes to visit a customer in Leeds then returns to Bristol. Stan knows that his car travels about 30 miles on each gallon of petrol. Find, to the nearest gallon, the amount of petrol Stan would expect to use on this day.

  A 7 gallons   **B** 11 gallons   **C** 14 gallons   **D** 15 gallons

Stan's company pays him for travelling expenses. He is allowed £27.50 per month plus 12.5 pence per mile he travels. Last month he travelled 6841 miles and received £882.63 in travelling expenses

3 Which of these calculations could Stan use to carry out a rough check on the amount he was paid for last month's travelling expenses?

  A $30 + 7000 \times 10$

  B $\dfrac{30 + 7000 \times 10}{100}$

  C $\dfrac{(30 + 10) \times 7000}{100}$

  D $30 + \dfrac{7000 \times 10}{100}$

Often questions involve checking. You are sometimes asked to select a method rather than working out a numerical value.

Here's how...

1 Distance from Bristol to Manchester = 172 miles

$$\text{Time} = \frac{\text{Distance}}{\text{Speed}} = \frac{172}{50} \approx 3\tfrac{1}{2} \text{ (hours)} \qquad \textbf{Answer} = C$$

> **checkpoint**
>
> $172 \div 50 = 17.2 \div 5 =$
>
> $$\begin{array}{r} 3\ .4\ 4 \\ 5\overline{)1\ 7\ .^2 2^2 0} \end{array}$$
>
> $3\tfrac{1}{2} = 3.5$ is the nearest option.

2 Distance from Bristol to Leeds = 213 miles

Total Distance $= 2 \times 213 = 426$ miles

Estimate of petrol used $= \dfrac{426}{30} \approx 14$ gallons $\qquad \textbf{Answer} = C$

> **checkpoint**
>
> $426 \div 30 = 42.6 \div 3 =$
>
> $$\begin{array}{r} 1\ 4\ .\ 2 \\ 3\overline{)4^1 2\ .^1 6} \end{array}$$
>
> 14 gallons is the nearest option.

3 Rounding values to 1 significant figure:

Stan is paid approximately £30 plus 10 pence per mile

7000 miles at 10 pence per mile amounts to 7000×10 **pence**

This amount must be divided by 100 to change it to £ before adding.

Rough estimate for expenses $= 30 + \dfrac{7000 \times 10}{100}$ $\qquad \textbf{Answer} = D$

> **checkpoint**
>
> Working out this estimate:
>
> $30 + \dfrac{7000 \times 10}{100} = 30 + 700 = £730$
>
> The other estimates give:
>
> A $30 + 7000 \times 10 = 30 + 70\,000 = £70\,030$
>
> B $\dfrac{30 + 7000 \times 10}{100} = \dfrac{70\,030}{100} = £700.30$
>
> C $\dfrac{(30 + 10) \times 7000}{100} = \dfrac{40 \times 7000}{100} = £2800$
>
> (If you have to guess in this case, do not choose A or C!)

Population

The table gives population figures in millions for the major regions of the world.

Area	Year					
	1800	1850	1900	1950	1998	2050
Africa	107	111	133	221	749	1766
Asia	635	809	947	1402	3585	5268
Europe	203	276	408	547	729	628
Latin America	24	38	74	167	504	809
North America	7	26	82	172	305	392
Oceania	2	2	6	13	30	46
World (Total)	978	1262	1650	2521	5901	8909

Table 5.1   Source: United Nations Website www.undp.org/popin/wdtrends

> **hint**
>
> The figures for 2050 are projected estimates

1 By approximately what percentage did the population of Europe increase in the 19th century (i.e. from 1800 to 1900)?

   A 25%   **B** 50%   **C** 100%   **D** 200%

> **hint**
>
> 'Approximately' indicates you should estimate the answer.

2 Which of these fractions is the best estimate for the fraction of the world's population living in Asia in 1998?

   A $\frac{1}{2}$ **B** $\frac{3}{5}$ **C** $\frac{2}{3}$ **D** $\frac{7}{10}$

3 Which of these is the most effective way to show changes in the total world population from 1800 to 2050?

A line graph   **B** scattergraph

C comparative bar-chart   **D** pie-chart

> **hint**
> Note that it is an effective illustration of changes in the *total* world population that is required, rather than the populations of the individual regions.

4 A pie-chart can be drawn to illustrate the proportion of the world's population expected to live in each region in 2050. Which of these calculations would give the angle used to represent Latin America?

A $\dfrac{360}{809} \times 8909$ **B** $\dfrac{809}{8909} \times 360$

C $\dfrac{8909}{809 \times 360}$ **D** $\dfrac{8909}{360} \times 809$

> **hint**
> This is another question asking for a method, rather than a value.

Here's how...

1 The population of Europe increased from 203 million to 408 million between 1800 and 1900. The increase was 205 million.

$$\% \text{ increase} = \frac{\text{increase}}{\text{original}} \times 100\% = \frac{205}{203} \times 100$$

$$\approx 1 \times 100 = 100\% \qquad \textbf{Answer} = \text{C}$$

> **hint**
> See Section 2.1.3 for percentage change.

2 3585 out of 5901 (million) lived in Asia in 1998

The fraction is $\dfrac{3585}{5901}$. Rounding to values that cancel easily:

$$\frac{3585}{5901} \approx \frac{3600}{6000} = \frac{36}{60} = \frac{6}{10} = \frac{3}{5} \qquad \textbf{Answer} = \text{B}$$

> **hint**
> Rounding 3585 to 4000 would give an estimate of $\frac{2}{3}$, but this is not as accurate. An alternative, but longer method is to try each option:
> A $\frac{1}{2}$ of 5900 = 2950
> B $\frac{3}{5}$ of 5900 = 3 × 1180 = 3540 (nearest to 3585)
> C $\frac{2}{3}$ of 5900 = 2 × 1967 = 3934
> D $\frac{7}{10}$ of 5900 = 7 × 590 = 4130

3 Neither a scatter graph nor a pie-chart are suitable for this data. In a comparative bar-chart the population in each region would be shown as bars alongside each other. The *total* population would not be obvious. A line graph is the most effective for showing changes in the total world population.

$$\textbf{Answer} = \text{A}$$

> **hint**
> See Section 3.1 for for charts and graphs.

4 809 million out of 8909 million are expected to live in Latin America in 2050.

This gives the fraction $\dfrac{809}{8909}$

The pie-chart angle is this fraction *of* 360°.

$$\text{Angle} = \frac{809}{8909} \times 360$$

$$\textbf{Answer} = \text{B}$$

> **checkpoint**
> You could use rough estimates to see which option is likely.
> e.g. A gives approximately
> $$\frac{400}{800} \times 9000 = 4500,$$
> obviously too big as a pie-chart has only 360°. (If you have to guess, guess sensibly!)

5.6 Sample test questions for level 3

Level 3 questions are set in real contexts and vary in length.
Section A contains relatively short questions like these:

Short Question   Depreciation of a car's value

a A new car costs £15 000 and depreciates in value by 12% each year.
Find the value of the car when it is 2 years old.   (2 marks)

b The value, £V, of a car when it is n years old is given by the formula:

$$V = P\left(1 - \frac{r}{100}\right)^n$$

where £P is its value when new and $r\%$ is the rate at which the value of the
car depreciates.
Find the rate at which a car depreciates if its value is reduced to
three-fifths of its original value in 3 years.   (3 marks)

hint

See Section 2.1.3 for repeated percentage changes. Losing 12% means the value at the end of each year is 88% of the value at the beginning of the year.

Here's how...

a At the end of 1st year, value of car in £ = 88% of 15 000

$$= 0.88 \times 15\ 000 = 13\ 200$$

At the end of 2nd year, value of car in £ = 88% of 13 200

$$= 0.88 \times 13\ 200 = 11\ 616$$

Answer: Value of the car when it is 2 years old is £11 616

checkpoint

Calculate 12% and subtract it at each stage.

b The car's value is reduced to three-fifths of its original value in 3 years.
The multiplier for three years must equal 0.6

$$\left(1 - \frac{r}{100}\right)^3 = 0.6$$

Take the cube root of both sides $1 - \dfrac{r}{100} = \sqrt[3]{0.6} = 0.8434...$

Rearrange the equation $\dfrac{r}{100} = 0.1566$

$$r = 15.66$$

Answer: The rate of depreciation is 16% per year (nearest %)

hint

$\frac{3}{5} = 0.6$

hint

Written out in full:

$1 - \dfrac{r}{100} = 0.8434$

$1 = 0.8434 + \dfrac{r}{100}$

$1 - 0.8434 = \dfrac{r}{100}$

$\dfrac{r}{100} = 0.1566$

hint

Round the answer sensibly.

Questions set in a real context may involve a variety of topics. You might be asked to explain how you would
check an answer. Sometimes it is possible to use a different method to do this, as in the next example. On
other occasions you might have to check the solution of an equation by checking that it fits the original
information given in the question. Occasionally you may need to use approximations or inverse methods.
Look at ToTT.8 if you need help with methods of checking.

Short Question   Swimming Pool

The sketch shows the cross section
of the water in a swimming pool.

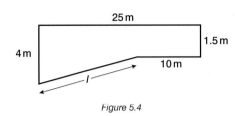

Figure 5.4

a Calculate, to the nearest tenth of
a metre, the length of the sloping
floor, *l*.   (2 marks)
b Calculate the angle made by
the sloping floor with the
horizontal.   (3 marks)
c Show how you could use a different method to check your
answer to part **b**   (2 marks)

The volume of water in the pool is 450 m³.

d If chlorine is added to the water at a rate of 8 ppm (parts per million), how
many litres of chlorine are required?   (1 mark)

Here's how...

a Add a line to form a
right-angled triangle.

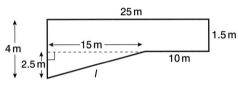

Figure 5.5

Using Pythagoras:
$$l^2 = 2.5^2 + 15^2$$
$$= 6.25 + 225$$
$$= 231.25$$
$$l = \sqrt{231.25} = 15.2069...$$

Answer: Length of the sloping floor is 15.2 m (1 d.p.)

b The angle required is θ.

Using

$$\tan\theta = \frac{o}{a} \text{ gives:}$$

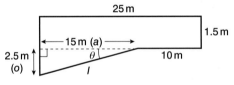

Figure 5.6

$$\tan\theta = \frac{2.5}{15} = 0.1\dot{6}$$
$$\theta = \tan^{-1} 0.1\dot{6}$$
$$= 9.4623...°$$

Answer: Angle between sloping floor and horizontal is 9.5° (1 d.p.)

c Using

$$\sin\theta = \frac{o}{h} \text{ gives:}$$

$$\sin\theta = \frac{2.5}{15.2069} = 0.16439...$$
$$\theta = \sin^{-1} 0.16439...$$
$$= 9.4623...°$$

Answer: Angle between sloping floor and horizontal is 9.5° (1 d.p.)

d Volume of water in litres = 450 × 1000 = 450 000 = 0.45 million litres
8 ppm means 8 litres of chlorine are needed for every million litres of water
Volume of chlorine required for the pool = 0.45 × 8 = 3.6 litres.

Answer: Volume of chlorine required is 3.6 litres (1 d.p.)

hint

See Section 2.3.4
Working with
right-angled triangles.

hint

Note that a *different*
method is required.
A rough check or inverse
of the *same* method will
not do.

hint

Two sides of the triangle
are found from the
original information:
25 − 10 = 15 and
4 − 1.5 = 2.5

hint

Write down 5 or 6 of the
figures given by your
calculator for *l* before
rounding. This enables
you to use a more
accurate value for *l* in
subsequent calculations.

hint

Avoid using the value
calculated for *l*. It is more
likely to contain an error.

hint

Round the answer
sensibly. 9° (nearest
degree) is also
acceptable.

hint

Do not use the rounded
value of *l* here. Using
more figures gives a
more accurate result.

hint

1 million = 1 000 000
1m³ = 1000 litres

The last question in Section B will be a long question, worth up to 25 of the 50 marks on the paper. This question will require you to draw at least one graph or chart. The question may involve statistics (as in the example below) or other topics.

Extended Question   Quick reactions

David thinks that playing computer games helps people to react more quickly in other situations. To test this theory he carries out an experiment with the help of the students in his class. Half of the students agree to play computer games for at least 10 hours in a particular week. The other students play no computer games at all. At the end of the week he measures how quickly each student reacts, using the method given below.

David holds the end of a centimetre ruler. The ruler hangs vertically down with the '0 cm' mark at the bottom. Without touching the ruler, a student, Ann, places her hand around the '0 cm' mark.

Figure 5.7

David releases the ruler without warning. Ann must close her hand as quickly as possible to catch the ruler as it falls. The measurement on the centimetre scale against the top of Ann's thumb is called the 'catch distance' (c.d.). The catch distance '7.3 cm' is used to represent Ann's reaction time.

Figure 5.8

Each student is tested in this way. The experimental results are given below:

Group A: played computer games				Group B: no computer games		
Name	Male/ Female	Time on games (hours)	c.d. (cm)	Name	Male/ Female	c.d. (cm)
Ann	F	15	7.3	Ali	M	6.3
Beth	F	18	5.9	Carol	F	8.9
Farid	M	13	8.5	Dan	M	9.5
Janet	F	10	9.3	Imran	M	7.9
John	M	12.5	8.2	Kay	F	10.8
Liam	M	10	9.8	Max	M	8.3
Sally	F	14.5	7.9	Neil	M	9.6
Tanya	F	17	6.5	Pat	F	6.3
Will	M	16	6.7	Zoe	F	10.7

Table 5.2

a Find the mean catch distance of the students who played computer games. (1 mark)

hint
See Section 3.3.1 for averages.

b The mean catch distance for students who did not play computer games was 8.7 cm. Do the results of the experiment support David's theory? Explain your answer. (1 mark)

c Which student in Group B had the median catch distance for that group? (2 marks)

d Find the modal catch distance for Group B. Explain why this is not an appropriate statistic to use as a representative value for the results of this group. (2 marks)

hint
See Section 3.4.4 for scatter graphs.

e Draw a scatter graph of catch distance against time spent on computer games for Group A. (6 marks)

f By inspection, draw a line of best fit on the scatter graph. (1 mark)

g Use your line of best fit to estimate the number of extra hours spent on computer games that would generally give a reduction of 1 cm in the catch distance. (2 marks)

h Explain why it might not be sensible to use the scatter graph to estimate the catch distance for a student who spent 24 hours on computer games during the week prior to doing the experiment. (1 mark)

A formula that approximately relates catch distance and reaction time is:

hint
See Section 2.2.1 for formulae and substitution.

$$T = \sqrt{\frac{d}{500}}$$

where T is the reaction time, in seconds, for a catch distance of d centimetres.

i Calculate the range of reaction times for each group. (5 marks)

hint
See Section 3.3.2 for range.

j To what level of accuracy would your answers to part **i** have to be given to show a difference between the range of reaction times for Group A and the range of reaction times for Group B? (1 mark)

hint
See Section 2.2.6 for rearranging formulae.

k Rearrange the formula $T = \sqrt{\dfrac{d}{500}}$ to make d the subject. (1 mark)

l In Group A, which student's reaction time was nearest to 0.128 seconds? (2 marks)

(Total 25 marks)

Here's how...

a Mean = $\dfrac{7.3 + 5.9 + 8.5 + 9.3 + 8.2 + 9.8 + 7.9 + 6.5 + 6.7}{9} = \dfrac{70.1}{9} = 7.7\dot{8}$

Mean catch distance for students who played games = **7.8 cm (1 d.p.)**

hint

Mean = $\dfrac{\text{sum of values}}{\text{number of values}}$

Round the answer sensibly.

b Yes, the results support David's theory because on average the students who have played computer games have a shorter catch distance. This means they are reacting more quickly.

c Putting the Group B results in order gives:

6.3 6.3 7.9 8.3 8.9 9.5 9.6 10.7 10.8
median

The student with the median catch distance in Group B is **Carol**.

hint

Median = middle value in ordered list.

d Modal catch distance for Group B = **6.3 cm**

It is not a good representative value because it is the lowest catch distance in the group.

hint

Mode = value that occurs most frequently.

e and f

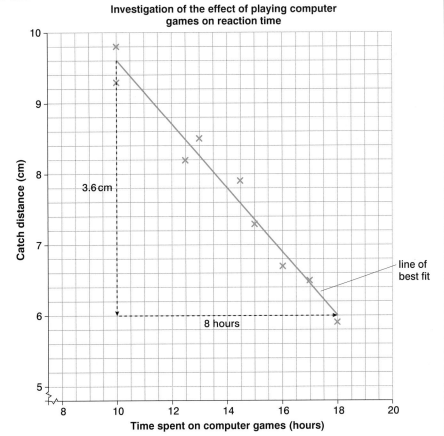

Figure 5.9

hint

Each square on the vertical scale represents 0.2 cm

hint

The line of best fit follows the general direction of the points. There are approximately the same number of points above and below it.

g Using the dotted lines:

Reduction of 3.6 cm results from an extra 8 hours of computer games.

Reduction of 1 cm results from: $\dfrac{8}{3.6}$ hours = $2.\dot{2}$ hours

To give a reduction of 1 cm would require approximately **2.2 hours** (1 d.p.)

hint

See Section 2.1.2 for reducing to unity.

hint

Round sensibly. As the answer is only approximate, 2 hours is acceptable.

h 24 hours is outside the range of experimental results.

hint

It is unlikely that catch distances can continue to reduce at the rate shown by the line of best fit – there is a limit to how quickly we can react.

i In Group A the shortest catch distance is 5.9 cm and the longest catch distance is 9.8 cm.

Substituting these values into $T = \sqrt{\dfrac{d}{500}}$ will give the shortest and longest reaction times in Group A:

hint

The longer the catch distance, the longer the reaction time.

Shortest reaction time in Group A = $\sqrt{\dfrac{5.9}{500}} = \sqrt{0.0118} = 0.1086...$ seconds.

hint

Enter this value into the calculator's memory for use later.

Longest reaction time in Group A = $\sqrt{\dfrac{9.8}{500}} = \sqrt{0.0196} = 0.14$ seconds.

hint

Range = highest value – lowest value where the lowest value is the value stored in the calculator's memory.

Range of reaction times for Group A = $0.14 - 0.1086... = 0.03137...$ sec.

Range of reaction times for Group A = 0.031 s (to 3 d.p.)

Shortest reaction time in Group B = $\sqrt{\dfrac{6.3}{500}} = \sqrt{0.0126} = 0.1122...$ seconds.

hint

Use the calculator's memory again.

Longest reaction time in Group B = $\sqrt{\dfrac{10.8}{500}} = \sqrt{0.0216} = 0.1469...$ seconds.

Range of reaction times for Group B = $0.1469... - 0.1122... = 0.0347...$ seconds

Range of reaction times for Group B = 0.035 s (to 3 d.p.)

j If both ranges found in part **i** were given to 2 decimal places this would not be accurate enough to show a difference since both ranges would then be 0.03 s. So the answers to part **i** must be given to **at least 3 decimal places**.

hint

'At least 2 significant figures' is an alternative answer.

k $\sqrt{\dfrac{d}{500}} = T$

hint

Turn the formula around

Square both sides: $\dfrac{d}{500} = T^2$

Multiply by 500: $d = 500T^2$

checkpoint

Substituting $d = 8.192$ into the formula for T should give 0.128.

l Substituting $T = 0.128$ into $d = 500T^2$ gives $d = 500 \times 0.128^2 = 8.192$ cm
The student in Group A whose catch distance is nearest to this value is **John**.

6 Tricks of the Trade

ToTT.1 Doing without a calculator

Addition

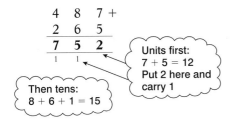

```
  4  8  7 +
  2  6  5
  7  5  2
  1  1
```

Then tens:
8 + 6 + 1 = 15

Units first:
7 + 5 = 12
Put 2 here and carry 1

Subtraction

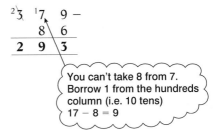

```
  ²3̷  ¹7̷  9 −
      8  6
  2  9  3
```

You can't take 8 from 7.
Borrow 1 from the hundreds column (i.e. 10 tens)
17 − 8 = 9

Multiplication by 10, 100, 1000, …

Add 1, 2, 3, … zeros to the end e.g. $354 \times 1000 = 354\,000$

Division by 10, 100, 1000, …

Remove 1, 2, 3, … zeros from the end e.g. $350\,000 \div 100 = 3500$
If there are not enough zeros, use decimals (ToTT.3).

Short multiplication

```
     5  7 ×
        6
  3  4  2
     4
```

6 times 7 is 42
Carry the 4, write in the 2
6 times 5 is 30 then add the 4

Long multiplication

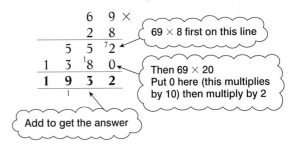

```
        6  9 ×
        2  8
     5  5  ⁷2
  1  3  ¹8  0
  1  9  3  2
        1
```

69 × 8 first on this line

Then 69 × 20
Put 0 here (this multiplies by 10) then multiply by 2

Add to get the answer

Short division

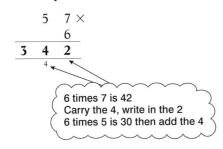

```
        7  4
  8)5  9  ³2
```

Seven 8s are 56 leaving 3 to carry

Four 8s make 32

Long division

```
           1  3  5
   14)1  8  9  0
      1  4
         4  9   Bring down the next figure, 9
         4  2   Three 14s are 42 leaving 7
            7  0   Bring down the next figure, 0
            7  0   Five 14s are 70 (none left)
```

One 14 from 18 leaves 4

It helps to try out multiples of 14

```
  1  4 ×        1  4 ×
     3             5
  4  2          7  0
  1             2
```

NB Don't rush to use these standard methods – there are sometimes much quicker ways!
E.g. $38 + 99 = 38 + 100 - 1 = 137$ is quicker than using addition method above.

ToTT.2 Fractions

Forming fractions

A **fraction** is part of a whole. For example $\frac{4}{9}$ means 4 parts out of 9. The **numerator** is 4 and the **denominator** is 9. The fraction can also be written on one line: $4/9$.

If necessary, convert the 'part' and the 'whole' to the same units. E.g. to express 7 minutes as a fraction of 2 hours, write 2 hours as 120 minutes. The required fraction is $\frac{7}{120}$.

The fractions of a whole must add up to 1

E.g. if $\frac{2}{5}$ of the tables in a restaurant are for 'smokers' then $\frac{3}{5}$ of the tables are for 'non-smokers' because $\frac{2}{5} + \frac{3}{5} = 1$ (see below for adding fractions).

Equivalent fractions

Two fractions that look different may be equal e.g. $\frac{2}{4}$ is the same as $\frac{1}{2}$ because $\frac{1}{2} = \frac{1 \times 2}{2 \times 2} = \frac{2}{4}$. We say $\frac{2}{4}$ is **equivalent** to $\frac{1}{2}$.

$\frac{6}{15}$ is equivalent to $\frac{2}{5}$ (multiply the numerator and denominator of $\frac{2}{5}$ by 3 to get $\frac{6}{15}$

Simplest terms

A fraction can be simplified by dividing its numerator *and* denominator by a suitable number.

E.g. $\frac{6}{15}$ can be simplified to $\frac{2}{5}$ by dividing the numerator and denominator by 3.

Since no number (apart from 1) exactly divides into both 2 *and* 5, $\frac{2}{5}$ is in its **simplest** terms.

Adding and subtracting fractions with the same denominators

To add or subtract fractions which have the same denominator, add or subtract the numerators.

E.g. $\frac{2}{7} + \frac{3}{7} = \frac{2+3}{7} = \frac{5}{7}$ and $\frac{7}{11} - \frac{4}{11} = \frac{7-4}{11} = \frac{3}{11}$

Adding and subtracting fractions with different denominators

Use a common denominator.

E.g. $\frac{2}{3} + \frac{1}{4} = \frac{8}{12} + \frac{3}{12} = \frac{11}{12}$ and $\frac{3}{4} - \frac{2}{5} = \frac{15}{20} - \frac{8}{20} = \frac{7}{20}$

Multiplying fractions together

Multiply the numerators and denominators separately, e.g. $\frac{2}{3} \times \frac{4}{5} = \frac{2 \times 4}{3 \times 5} = \frac{8}{15}$

The arithmetic is sometimes simplified by cancelling: $\frac{\overset{5}{\cancel{15}}}{\underset{7}{\cancel{28}}} \times \frac{\overset{1}{\cancel{4}}}{\underset{3}{\cancel{9}}} = \frac{5}{21}$

Dividing one fraction by another

Turn the fraction after the '÷' sign upside down (*invert*). Then multiply the fractions together.

E.g. $\frac{3}{7} \div \frac{4}{5} = \frac{3}{7} \times \frac{5}{4} = \frac{15}{28}$

Improper fractions and mixed numbers

$\frac{4}{3}$ is an **improper** fraction because its numerator is bigger than its denominator. $3\frac{1}{2}$ is a mixed number. It has a whole number (3) and a fraction ($\frac{1}{2}$). Every mixed number can be changed into an improper fraction e.g. to change $2\frac{3}{4}$ into an improper fraction...

$$2\frac{3}{4} = 2 + \frac{3}{4} = \frac{8}{4} + \frac{3}{4} = \frac{11}{4}$$

> 2 can be written as $\frac{8}{4}$

To change an improper fraction into a mixed number, divide and find the remainder.

E.g. to express $\frac{11}{5}$ as a mixed number $5\overline{)11}^{\,2r1}$ and so $\frac{11}{5} = 2\frac{1}{5}$

ToTT.3 Decimals

Decimals can be written as fractions

Put the figures from the decimal in the numerator. Put the place value of the **last** digit in the denominator.

Decimals			Fraction	Fraction in Lowest Terms
place value	$\frac{1}{10}$ $\frac{1}{100}$ $\frac{1}{1000}$			
0 . 7			$\frac{7}{10}$	$\frac{7}{10}$ (does not cancel)
0 . 0 0 2			$\frac{2}{1000}$	$\frac{1}{500}$ (cancelled by 2)
0 . 6 5			$\frac{65}{100}$	$\frac{13}{20}$ (cancelled by 5)
0 . 0 1 6			$\frac{16}{1000}$	$\frac{2}{125}$ (cancelled by 8)

Fractions can be written as decimals

Divide the numerator by the denominator.
Add zeros if if necessary.

$$\begin{array}{r} 0\ .\ 6\ 2\ 5 \\ 8\overline{)5\ .\ {}^50\ {}^20\ {}^40} \end{array}$$

$\frac{5}{8} = 0.625$

Sometimes fractions give *recurring decimals*.

e.g. $\frac{2}{9} = 0.2222...$ is written as $0.\dot{2}$

$$\begin{array}{r} 0\ .\ 2\ 2\ 2 \\ 9\overline{)2\ .\ {}^20\ {}^20\ {}^20} \end{array}$$

This division never ends

A group of recurring digits is written with a dot over the first and last recurring digit.

Decimals can be added or subtracted

Keep the decimal points in line.

e.g. $3.54 + 0.267 + 42.8 = \mathbf{46.607}$

$$\begin{array}{r} 3\ .\ 5\ 4\ \ + \\ 0\ .\ 2\ 6\ 7 \\ 4\ 2\ .\ 8 \\ \hline 4\ 6\ .\ 6\ 0\ 7 \\ {}_1\quad\ {}_1 \end{array}$$

When subtracting, add zeros if necessary.

e.g. $34.5 - 6.38 = \mathbf{28.12}$

$$\begin{array}{r} {}^2\cancel{3}\ {}^1 4\ .\ {}^4\cancel{5}\ {}^10\ \ - \\ 6\ .\ 3\ 8 \\ \hline 2\ 8\ .\ 1\ 2 \end{array}$$

Extra zero here

Decimals can be multiplied

- Ignore the decimal points and multiply the numbers as if they were whole numbers.
- Count the total number of decimal digits in the numbers being multiplied.
- Put the decimal point in the answer by counting from the end. Insert zeros if necessary.
- Insert zeros if necessary.

e.g. $1.3 \times 2 = \mathbf{2.6}$ (1 decimal digit) $0.1 \times 0.3 = \mathbf{0.03}$ (2 decimal digits)

$0.05 \times 1.2 = 0.060 = \mathbf{0.06}$ (3 decimal digits at first, but the final zero can then be omitted)

To multiply by 10, 100, 1000, … move the decimal point 1, 2, 3, … places to the right.

e.g. $0.35 \times 10 = 3.5$ and $2.75 \times 1000 = 2750$

Splitting the multiplication sometimes helps.

e.g. To multiply by 30 multiply by 10 then by 3: $0.015 \times 30 = 0.15 \times 3 = 0.45$

Multiplying is more difficult with harder numbers.

38×0.07

Start by ignoring the decimal points:

```
  3  8 ×
     7
  2  6  6
     5
```

There are 2 decimal digits in 38×0.07 so the answer is **2.66**

0.49×3.6

```
     4  9 ×
     3  6
     2  9  4
  1  4  7  0
  1  7  6  4
        1
```

There are 3 decimal digits in 0.49×3.6 so the answer is **1.764**

Decimals can be divided

You may need to add zeros e.g. $9.2 \div 8 = \mathbf{1.15}$

```
      1 .  1  5
  8)9 . ¹2 ⁴0
```

Extra zero here

To divide by 10, 100, 1000, … move the decimal point 1, 2, 3, … places to the left.

e.g. $8.7 \div 10 = 0.87$ and $7.53 \div 100 = 0.0753$

Splitting the division sometimes helps.

e.g. To work out $4.5 \div 500$ divide by 100 first: $4.5 \div 100 = 0.045$ then by 5: $0.045 \div 5 = 0.009$

- **To divide by a decimal, change it into a whole number (by multiplying it by 10 or 100 or 1000 …).**
- **To get the correct answer you must also change the other number in the same way.**

e.g. $0.0028 \div 0.04 = 0.28 \div 4 = \mathbf{0.07}$ (Each number is multiplied by 100 to make 0.04 into a whole number.)

More difficult numbers may require long division.
The example shows $8.26 \div 1.4 = 82.6 \div 14 = 5.9$
(The divisor is made into a whole number $\times 10$)

```
           5 .  9
  14)8  2 .  6
     7  0
     1  2  6
     1  2  6
```

Rough Work

```
  1  4 ×          1  4
     5               9
  7  0            1  2  6
  2                3
```

ToTT.4 Ratios

Ratios and parts

A ratio describes how parts of a quantity compare with each other. Suppose that in an election for every two people who voted for A, there were three people who voted for B. The **ratio** of votes A to B is 2:3. Alternatively, the ratio of votes B to A is 3:2.

Equivalent ratios

Two ratios may use different numbers but be equal to each other.

E.g. the ratios 1:2 and 2:4 are equivalent because $1:2 = 1 \times 2 : 2 \times 2 = 2:4$

Similarly, 12:21 is equivalent to 4:7   *multiply each number in 4:7 by 3*

Simplest terms

Divide the parts by a suitable number. Keep going until nothing will divide both parts.
E.g. the ratio 18:24 can be simplified as follows:

Each part can be exactly divided by 2:   [÷2]   18:24 = 9:12

9 and 12 can both be exactly divided by 3:   [÷3]   9:24 = 3:4

Nothing (apart from 1) divides both 3 and 4 so 3:4 is in its **simplest** terms. Alternatively, the answer 3:4 can be reached immediately by dividing 18 and 24 by 6.

A ratio may involve more than two parts e.g. to express 16:24:32 in its simplest terms, divide each part by 8:
[÷8]   16:24:32 = 2:3:4.

Ratios with fractional parts

Whole numbers are easier to work with than fractions. The ratio $\frac{1}{2}:\frac{1}{3}$ can be expressed as a ratio with whole numbers by multiplying each part by $2 \times 3 = 6$:

$$[\times 6] \quad \tfrac{1}{2}:\tfrac{1}{3} = \tfrac{1}{2} \times 6 : \tfrac{1}{3} \times 6 = 3:2$$

Similarly, to express $\frac{5}{6}:\frac{3}{8}$ as a ratio with whole numbers multiply each part by $6 \times 8 = 48$

To simplify the arithmetic, you could multiply by 24 instead of 48. This works because both 6 and 8 divide exactly into 24. Whichever number you use, the answer in its simplest terms is 20:9

Ratios with decimal parts

Multiply by 10, 100, etc... to remove the decimal point. E.g. the decimal ratio 1.3:3.4 can be expressed as a whole number ratio by multiplying each part by 10:

$$[\times 10] \quad 1.3:3.4 = 13:34$$

The part with the most decimal places determines which number (10, 100, etc...) is needed. E.g. if the decimal ratio is 1.4:0.08 then multiply each part by 100 to turn 0.08 into a whole number.

$$[\times 100] \quad 1.4:0.08 = 140:8 = 35:2 \quad \textit{simplest terms}$$

Ratios and units

A ratio only makes sense if the parts are measured in the same units. E.g. to express £2.45:28p in its simplest terms, first convert £2.45 into pence: £2.45 = 245p
The ratio (in pence) is 245:28   *or 35:4 in its simplest terms*

Ratios and fractions

The smaller of the two parts can be expressed as a fraction of the larger part e.g. the ratio 3:4 means the smaller part is $\frac{3}{4}$ of the larger part.

One part of a ratio can be written as a fraction of the whole by adding the parts. E.g. if the ratio of males to females at a conference is 2:3 then the fraction of people which are male is 2 parts out of $5 (= 2 + 3)$ i.e. $\frac{2}{5}$

A fraction of a total gives the ratio for the parts. Suppose $\frac{5}{9}$ of a group pass their driving test. Then in a typical group of 9 learners, 5 will pass and the remaining 4 will fail. The 'pass:fail' ratio is 5:4.

ToTT.5 Percentages

Percentages to fractions

A percentage is a fraction with denominator 100. For example, $10\% = \frac{10}{100}$ and stands for '10 parts out of every 100'. A percentage may be written as a fraction in its simplest terms.

E.g. $25\% = \frac{25}{100} = \frac{1}{4}$ and $85\% = \frac{85}{100} = \frac{17}{20}$

Fractions to percentages

100% represents the total. To turn a fraction of a total into a percentage, multiply 100% by the fraction e.g. to turn $\frac{3}{5}$ into a percentage, find $\frac{3}{5}$ of 100% as follows:

first divide 100 by 5 to find $\frac{1}{5}$: $100 \div 5 = 20$

then multiply by 3 to find $\frac{3}{5}$: $3 \times 20 = 60$ i.e. $\frac{3}{5} = 60\%$

To turn $\frac{3}{8}$ into a percentage: divide 100 by 8 to find $\frac{1}{8}$: $8\overline{)100}^{\,12.5}$

multiply by 3 to find $\frac{3}{8}$: $3 \times 12.5 = 37.5\%$

Writing one quantity as a percentage of another

First find the **fraction** then turn the fraction into a percentage. It often helps to express the fraction in its simplest terms first e.g. £36 as a fraction of £90 is $\frac{36}{90} = \frac{2}{5}$ (simplest terms)

As a percentage $\frac{2}{5} = 40\%$ (use techniques above) i.e. £36 as a percentage of £90 is 40%

The percentages of a total must add up to 100%

For example, suppose 15% of drivers in a survey use leaded petrol. The percentage of drivers in the survey who do **not** use leaded petrol is $100\% - 15\% = 85\%$

Finding a percentage of an amount

Percentages are fractions. Multiply the amount by the fraction e.g. to find 7% of £600 you must find $\frac{7}{100} \times 600$

divide 600 by 100 to find $\frac{1}{100}$: $600 \div 100 = 6$

then multiply by 7 to find $\frac{7}{100}$: $7 \times 6 = 42$ i.e. 7% of £600 = £42

Short cuts for some percentages

- Finding 10% of an amount: 10% is the same as '1 in every 10'. To work out 10%, just divide the amount by 10 to find one-tenth e.g. 10% of £120 = $120 \div 10 = $ **£12**
- Finding 20%, 30% etc… is easy once you know 10%
  10% of £120 = **£12**   20% of 120 = $2 \times 12 = $ **£24**   30% of 120 = $3 \times 12 = $ **£36**   etc…
- Finding 5% and $2\frac{1}{2}\%$ is also easy once you know 10%
  10% of £120 = **£12**   5% of 120 = $12 \div 2 = $ **£6** $2\frac{1}{2}\%$ of 120 = $6 \div 2 = $ **£3**   etc…

The shortcuts can be used to find more complicated percentages e.g. VAT is charged at $17\frac{1}{2}\%$. The VAT on £600 is worked out as follows:

first find 10%:   10% of £600 = $600 \div 10 = £60$

notice that $17\frac{1}{2}\% = 10\% + 5\% + 2\frac{1}{2}\%$

          $= £60 + £30 + £15$

          $= £105$ i.e. $17\frac{1}{2}\%$ of £600 = £105 in VAT

Decimals to percentages

Decimals between 0 and 1 can be written as percentages by multiplying the decimal by 100. The decimal point moves two places to the right. Add extra zeros if necessary. E.g. to turn 0.025 in a percentage, work out $0.025 \times 100 = 2.5\%$. Similarly, 0.5 as a percentage is $0.500 \times 100 = 50.0\%$ or simply 50%

Percentages to decimals

Divide the percentage by 100 – i.e. move the decimal point 2 places to the left.
E.g. $36.4\% = 36.4 \div 100 = 0.364$ i.e. $36.4\% = 0.364$.
For whole numbers, the decimal point is at the end of the number: $7\% = 7.0 \div 100 = 0.07$ i.e. $7\% = 0.07$

ToTT.6 Negative numbers

Numbers and signs

Negative numbers are used for quantities which are **less than** zero. A '−' sign in front of a number shows the number is negative. E.g. a temperature of −5°C means 5 degrees **below** zero.

Negative answers occur when you try to subtract more than you've got. If a bank account holds £200 and a withdrawal of £300 is made then the account holder is overdrawn by £100.

$$£200 - £300 = -£100$$

Just for the record, numbers greater than zero are called **positive** numbers. A '+' sign can be placed in front of a number to show that it is positive. However, we usually leave out the '+' sign.

Temperature provides another use of negative numbers. Suppose that a thermometer at midnight shows 2°C and that the temperature overnight falls by 5°C. The temperature in the morning is $2 - 5 = -3$°C. If the temperature during the morning then rises by 7°C then the mid-day temperature is $-3 + 7 = 4$°C.

Combining adjacent signs

Adjacent signs are signs next to each other. E.g. in the sum $5 + -2$ the addition sign + is next to the negative sign −. The two rules for dealing with adjacent signs are:

Rule	Example
1 **Opposite signs** next to each other means **subtract**	$5 + -2 = 5 - 2$ $\qquad\quad = 3$
2 The same signs next to each other means **add**	$5 - -2 = 5 + 2$ $\qquad\quad = 7$

E.g. $\qquad 6 + -9 = 6 - 9$ rule 1
$$\qquad\qquad\quad = -3$$

and $\quad 20 + 6 - -8 = 20 + 6 + 8$ rule 2
$$\qquad\qquad\qquad = 20 + 14$$
$$\qquad\qquad\qquad = 34$$

Multiplying and dividing with signs

The rules for multiplying and dividing with signs are:

Rule	Example
1 Multiplying or dividing **different signs** gives a **negative** answer	$3 \times -2 = -6$ and $-12 \div 3 = -4$
2 Multiplying or dividing the **same signs** gives a **positive** answer	$3 \times 2 = 6$ and $-12 \div -3 = 4$

E.g. to work out the value of -4×3: first ignore the signs ($4 \times 3 = 12$). Since the signs are different the answer is negative i.e. $-4 \times 3 = -12$

Similarly, $\quad -2 \times -3 \times 4 = 6 \times 4$ $-2 \times -3 = 6$ by rule 2
$$\qquad\qquad\qquad\quad = 24$$

and $\qquad 5 \times -6 \div 3 = -30 \div 3$ $5 \times -6 = -30$ by rule 1
$$\qquad\qquad\qquad\quad = -10$$

Negative numbers and brackets

As a general rule, always work out the inside of the bracket first. E.g. to work out $2 \times (3 + -4)$:

deal with the bracket first: $\qquad 3 + -4 = 3 - 4 = -1$

then multiply by 2: $\qquad\qquad 2 \times -1 = -2 \qquad$ i.e. $2 \times (3 + -4) = -2$

Similarly, $\qquad\qquad\qquad -3 \times (2 - -4) = -3 \times 6$
$$\qquad\qquad\qquad\qquad\qquad = -18 \qquad \text{i.e. } -3 \times (2 - -4) = -18$$

ToTT.7 Using the calculator

Consult your instruction booklet or teacher if your calculator operates in a different way from that described. When you use a calculator ***always check the answer***. ToTT.8 gives help with this.

Errors can be corrected

The 'All-Clear' key $\boxed{\text{AC}}$ clears everything. The $\boxed{\text{C}}$ key clears just the last entry.

BODMAS gives the correct order for calculations

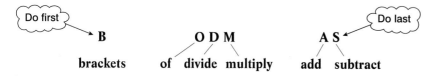

For example, dividing is done before adding, so $6 + 12 \div 3 = 6 + 4 = 10$ (***not*** $18 \div 3 = 6$)

Scientific calculators do calculations in the right order.

Press $\boxed{6}\boxed{+}\boxed{1}\boxed{2}\boxed{\div}\boxed{3}\boxed{=}$ to check your calculator gives 10.

Sometimes you need to put in brackets

The correct answer for $\dfrac{84 + 80}{80 - 76}$ is $\dfrac{164}{4} = 41$

but if you press $\boxed{8}\boxed{4}\boxed{+}\boxed{8}\boxed{0}\boxed{\div}\boxed{8}\boxed{0}\boxed{-}\boxed{7}\boxed{6}\boxed{=}$ you will get 9.

The calculator has used BODMAS and worked out $80 \div 80$ first.

The correct answer can be found on the calculator using brackets: $\dfrac{(84 + 80)}{(80 - 76)}$

Press $\boxed{(}\boxed{8}\boxed{4}\boxed{+}\boxed{8}\boxed{0}\boxed{)}\boxed{\div}\boxed{(}\boxed{8}\boxed{0}\boxed{-}\boxed{7}\boxed{6}\boxed{)}\boxed{=}$ to get the correct answer, **41**.

Numbers can be stored in the memory

$\boxed{\text{Min}}$ is used to put numbers into the memory. $\boxed{\text{MR}}$ is used to recall numbers from the memory.

The last example can be done by storing the value of the denominator in the memory:

Press $\boxed{8}\boxed{0}\boxed{-}\boxed{7}\boxed{6}\boxed{=}\boxed{\text{Min}}$ $\boxed{8}\boxed{4}\boxed{+}\boxed{8}\boxed{0}\boxed{=}\boxed{\div}\boxed{\text{MR}}\boxed{=}$ giving **41** again.

Negative numbers are entered into a calculator using the $\boxed{+/-}$ key

To calculate -2×-5 press $\boxed{2}\boxed{+/-}\boxed{\times}\boxed{5}\boxed{+/-}\boxed{=}$ The answer is **10**

Use the fraction key $a^b/_c$ to enter fractions and mixed numbers

	Example	Key in:	Display	Answer
Entering a fraction	$\frac{5}{8}$	[5] [$a^b/_c$] [8]	5⌐8 (or 5r8)	
Equivalent fractions –cancelling to simplest form	$\frac{6}{9}$	[6] [$a^b/_c$] [9] [=]	2⌐3	$\frac{2}{3}$
Improper fractions –changing to mixed numbers	$\frac{17}{3}$	[1] [7] [$a^b/_c$] [3] [=]	5⌐2⌐3	$5\frac{2}{3}$
Mixed numbers –changing to improper fractions	$3\frac{2}{5}$	[3] [$a^b/_c$] [2] [$a^b/_c$] [5] [$d/_c$]	17⌐5	$\frac{17}{5}$
Finding a fraction of a number –remember 'of' is ×	$\frac{3}{4}$ of 6	[3] [$a^b/_c$] [4] [×] [6] [=]	4⌐1⌐2	$4\frac{1}{2}$
Fraction arithmetic –use +, − ×, ÷ as usual	$\frac{5}{6} - \frac{4}{5}$	[5] [$a^b/_c$] [6] [−] [4] [$a^b/_c$] [5] [=]	1⌐30	$\frac{1}{30}$
Changing fraction to decimal –do as a division	$\frac{5}{16}$	[5] [÷] [1] [6] [=]	0.3125	0.3125
Changing decimal to fraction –write as a fraction yourself then cancel to simplest form	0.48	$\frac{48}{100}$ then using the calculator: [4] [8] [$a^b/_c$] [1] [0] [0] [=]	12⌐25	$\frac{12}{25}$

It is recommended that you use the methods given in ToTT.5 for percentages, rather than the % key.

There are keys for powers and roots

Use x^2 for squares, x^y for other powers $\sqrt{}$ for square roots and $x^{1/y}$ for other roots.

	Example	Key in:	Answer
Squaring a number	4^2 (means 4×4)	[4] [x^2]	16
Finding other powers	4^3 (means $4 \times 4 \times 4$)	[4] [x^y] [3] [=]	64
Taking the square root of a number	$\sqrt{25}$	[2] [5] [$\sqrt{}$] or [$\sqrt{}$] [2] [5]	5
Finding other roots	$\sqrt[3]{8}$	[8] [$\sqrt[3]{}$] or [8] [$x^{1/y}$] [3] [=]	2

LEVEL 3

Enter numbers in standard form using the [EXP] key

To enter 5×10^8 press [5] [EXP] [4] *NB DO NOT ENTER THE ×10*

Standard form is explained in Section 2.1.4

ToTT.8 Accuracy and checking

Measurements are never exact

The accuracy depends on the instrument used.
The amount of water in this measuring jug is between 700 and 750 m*l*.
A reasonable estimate is 730 m*l*.

Sometimes numbers are rounded

Suppose there are 16 394 females and 27 602 males at a football match.
To the nearest thousand there are 16 000 females and 28 000 males.
The number of females has been rounded down because 16 394 is less than
halfway between 16 000 and 17 000 (394 is less than 500).

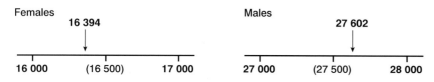

The number of males has been rounded up because 27 602 is more than halfway between 27 000 and 28 000
(602 is more than 500).

Examples

Number	Rounded to	Answer	Reason
654	nearest hundred	700	654 is nearer to 700 than 600 (because 54 is greater than 50)
82	nearest ten	80	82 is nearer to 80 than 90 (because 2 is less than 5)
3.7	nearest whole number	4	3.7 is nearer to 4 than 3 (because 0.7 is greater than 0.5)
21.5	nearest whole number	22	21.5 is *exactly* half-way between 21 and 22. Normally such numbers are rounded up, though 21 would be equally near.

You can round using decimal places

To round to **1** decimal place, look at the **2nd** decimal place.
If this is less than 5 ignore this digit and any following digits.
If the 2nd decimal place is 5 or more, round the 1st decimal place upwards then ignore all following digits.

Examples (NB d.p. means decimal place)

Number	Rounded to	Answer	Reason (the deciding digit is in bold)
2.3**7**	1 d.p.	2.4	7 is greater than 5 – round the first decimal place upwards
14.6**2**8	1 d.p.	14.6	2 is less than 5 – ignore this digit and any that follow it
0.27**6**	2 d.p.	0.28	6 is greater than 5 – round the second decimal place upwards
1.40**5**2	2 d.p.	1.41	The 3rd decimal place is 5 – round the second place upwards
0.029**8**	3 d.p.	0.030	8 is greater than 5 – round the third decimal place upwards. In this case 29 must be rounded up to 30. The last zero must be included. (0.03 has only 2 decimal places.)

You can round using significant figures

These numbers all have 2 significant figures: 31   3.1   0.31   0.031
The zeros in the last two numbers are there to give place value and do not count as significant figures.

Examples   **NB** s.f. means significant figure

Number	Rounded to	Answer	Reason (the deciding digit is in bold)
76.3**8**	3 s.f.	76.4	8 is greater than 5 – round the third significant figure upwards.
76.**3**8	2 s.f.	76	3 is less than 5 – ignore this digit and any that follow it.
76.38	1 s.f.	80	**6** is greater than 5 – round the first significant figure upwards to 8. The zero at the end of 80 is not significant – it is just there to maintain size. (Round to 80 because 76.38 is nearer to 80 than 70.)
0.92**4**9	2 s.f.	0.92	4 is less than 5 – ignore this digit and any that follow it. The zero before the point is not significant.
60 2**5**1	3 s.f.	60 300	When the deciding digit is 5, round up. The zero between 6 and 3 *is* significant but those at the end are not.
1.39**7**2	3 s.f.	1.40	7 is greater than 5 – the 39 must be rounded up to 40. Note that the zero at the end of this answer *is* significant because it contributes to the accuracy. (1.4 would be incorrect – it has only 2 significant figures.)

Measurements do not give precise answers in calculations

Answers should be rounded to the same degree of accuracy as the least accurate measurement used.

Always check calculations

Use one or more of the following methods:

- *Check that the answer makes sense*

- *Redo the calculation in a different way*
  For example, $37.63 - 24.84 + 16.3 = 29.09$ can be checked by doing the addition before the subtraction:
  $$37.63 + 16.3 - 24.84 = 29.09$$

- *Use estimates*
  For example, a calculator gives $\dfrac{392 \times 0.725}{6.4 + 8.23} = 19.4$ (to 3 s.f.)

  To check, round all figures to 1 s.f.:   Estimate $= \dfrac{400 \times 0.7}{6 + 8} = \dfrac{280}{14} = 20$

  Round to numbers that are easy enough to use without a calculator (may not always be to 1 s.f.)

- *Check using inverse operations*
  Addition and subtraction are inverses (opposites). Multiplication and division are inverses.
  Squaring and taking the square root are also inverses.
  Carrying out inverse operations in reverse order should always take you back to where you started.

  To check $37.63 - 24.84 + 16.3 = 29.09$ using inverse operations, work out $29.09 - 16.3 + 24.84$.
  The answer should be 37.63.

Part A What you need to know

Interpreting information

You need to know how to:

- obtain relevant information from different sources *(e.g. from written and graphical material, first-hand by measuring or observing)*
- read and understand graphs, tables, charts and diagrams *(e.g. frequency diagrams)*
- read and understand numbers used in different ways, including negative numbers *(e.g. for losses in trading, low temperatures)*
- estimate amounts and proportions
- read scales on a range of equipment to given levels of accuracy *(e.g. to the nearest 10 mm or nearest inch)*
- make accurate observations *(e.g. count the number of customers per hour)*
- select appropriate methods for obtaining the results you need, including grouping data when this is appropriate *(e.g. heights, salary bands)*.

Carrying out calculations

You need to know how to:

- show clearly your methods of carrying out calculations and give the level of accuracy of your results
- carry out calculations involving two or more steps, with numbers of any size
- convert between fractions, decimals and percentages
- convert measurements between systems *(e.g. from pounds to kilograms, between currencies)*
- work out areas and volumes *(e.g. area of an L-shaped room, number of containers to fill a given space)*
- work out dimensions from scale drawings *(e.g. using a 1:20 scale)*
- use proportion and calculate using ratios where appropriate
- compare sets of data with a minimum of 20 items *(e.g. using percentages, using mean, median, mode)*
- use range to describe the spread within sets of data
- understand and use given formulae *(e.g. for calculating volumes, areas of circles, insurance premiums, $V = IR$ for electricity)*
- check your methods in ways that pick up faults and make sure your results make sense.

Interpreting results and presenting your findings

You need to know how to:

- select effective ways to present your findings
- construct and use graphs, charts and diagrams *(e.g. pie-charts, frequency tables, workshop drawings)*, and follow accepted conventions for labelling these *(e.g. appropriate scales and axes)*
- highlight the main points of your findings and describe your methods
- explain how the results of calculations meet the purpose of your activity.

Part B What you must do

You must

Carry through at least one substantial activity that includes straightforward tasks for N2.1, N2.2 and N2.3

N2.1

Interpret information from **two** different sources, including material containing a graph.

Evidence must show you can:

- choose how to obtain the information needed to meet the purpose of your activity
- obtain the relevant information
- select appropriate methods to get the results you need.

N2.2

Carry out calculations to do with:

a amounts and sizes
b scales and proportion
c handling statistics
d using formulae.

- carry out calculations, clearly showing your methods and levels of accuracy
- check your methods to identify and correct any errors, and make sure your results make sense.

N2.3

Interpret the results of your calculations and present your findings. You must use at least **one** graph, **one** chart and **one** diagram.

- select effective ways to present your findings
- present your findings clearly and describe your methods
- explain how the results of your calculations meet the purpose of your activity.

Part A What you need to know

Planning an activity and interpreting information

You need to know how to:

- plan a substantial and complex activity by breaking it down into a series of tasks
- obtain relevant information from different sources, including a large data set **(over 50 items)**, and use this to meet the purpose of your activity
- use estimation to help you plan, multiplying and dividing numbers of any size rounded to one significant figure
- make accurate and reliable observations over time and use suitable equipment to measure in a variety of appropriate units
- read and understand scale drawings, graphs, complex tables and charts
- read and understand ways of writing very large and very small numbers *(e.g. £1.5 billion, 2.4 × 10⁻³)*
- understand and use compound measures *(e.g. speed in kph, pressures in psi, concentrations in ppm)*
- choose appropriate methods for obtaining the results you need and justify your choice.

Carrying out calculations

You need to know how to:

- show your methods clearly and work to appropriate levels of accuracy
- carry out multi-stage calculations with numbers of any size *(e.g. find the results of growth at 8% over three years, find the volume of water in a swimming pool)*
- use powers and roots *(e.g. work out interest on £5000 at 5% over three years)*
- work out missing angles and sides in right-angled triangles from known sides and angles
- work out proportional change *(e.g. add VAT at 17.5% by multiplying by 1.175)*
- work out actual measurements from scale drawings *(e.g. room or site plan, map, workshop drawing)* and scale quantities up and down
- work with large data sets **(over 50 items)**, using measures of average and range to compare distributions, and estimate mean, median and range of grouped data
- re-arrange and use formulae, equations and expressions *(e.g. formulae in spreadsheets, finance, and area and volume calculations)*
- use checking procedures to identify errors in methods and results.

Interpreting results and presenting your findings

You need to know how to:

- select and use appropriate methods to illustrate findings, show trends and make comparisons
- examine critically, and justify, your choice of methods
- construct and label charts, graphs, diagrams and scale drawings using accepted conventions
- draw appropriate conclusions based on your findings, including how possible sources of error might have affected your results
- explain how your results relate to the purpose of your activity.

Part B What you must do

You must

Plan and carry through at least one substantial and complex activity that includes tasks for N3.1, N3.2 and N3.3

N3.1

Plan and interpret information from **two** different types of sources, including a large data set.

Evidence must show you can:

- plan how to obtain and use the information required to meet the purpose of your activity
- obtain the relevant information
- choose appropriate methods for obtaining the results you need and justify your choice.

N3.2

Carry out multi-stage calculations to do with:

a amounts and sizes;

b scales and proportion;

c handling statistics;

d rearranging and using formulae.

You should work with a large data set on at least **one** occasion.

- carry out calculations, to appropriate levels of accuracy, clearly showing your methods
- check methods and results to help ensure errors are found and corrected.

N3.3

Interpret results of your calculations, present your findings and justify your methods. You must use at least **one** graph, **one** chart and **one** diagram.

- select appropriate methods of presentation and justify your choice
- present your findings effectively
- explain how the results of your calculations relate to the purpose of your activity.

Answers

NB Full answers can be found on the website www.nelsonthornes.com/gofigure.

2.1 Calculation Techniques

2.1.1 Working with fractions

'of' means multiply

1 a £6   b £8   c £140   d £15
   e £4.50   f £75   g £7.50   h £2.25
   i £4.20   j £13.33
2 640
3 48 million
4 35 minutes
5 a £315 000   b £525 000
6 60

Calculations may involve several steps

1 a $\frac{9}{10}$ b $\frac{15}{28}$ c $\frac{17}{45}$ d $\frac{1}{8}$
2 10
3 20
4 4
5 1

Fraction of a fraction

1 a $\frac{6}{35}$ b $\frac{1}{18}$ c $\frac{18}{25}$
2 a 18   b 10   c 25
3 a 15   a 12
4 8
5 a 72 000   b 18 000
6 21
7 12 times
8 a 17 fence posts   b 48 planks
9 a 168
   b

		Exam 1	
		Pass	Fail
Exam 2	Pass	$\frac{3}{5}$	$\frac{1}{20}$
	Fail	$\frac{3}{20}$	$\frac{1}{5}$

   c 42   d 196

An amount can be increased or decreased by a fraction

1 a i £16   ii £66   iii £50   iv £31.50
   b i £24   ii £30   iii £84
2 £240
3 68 000
4 8 deaths per thousand
5 a £2600   b £650
6 a 30   b i true   ii false
7 a 1.9 million   b No
      More tested or better diagnosis techniques

8 a 80
   b

Year	1998	1999	2000
RA deaths	80	96 $(+\frac{1}{5})$	132 $(+\frac{3}{8})$
Total No. of deaths	600	540 $(-\frac{1}{10})$	480 $(-\frac{1}{9})$

   c Proportion in 2000 $= \frac{132}{480} = \frac{33}{120}$
      Double 1998 proportion $= \frac{4}{15} = \frac{32}{120}$
      The expert is correct.

2.1.2 Working with ratios and proportion

An amount can be divided in a given ratio

1 a £16:£24   b £60:£100
   c £400:£600:£1400
2 a 6 calls   b 15 calls
3 a £630   b £1050
4 a 25 orange   b 20 mango
5 C (Must be a multiple of 5 and also of 7)
6 a 10
   b

	Male	Female	Total
Colour blind	4	6	10
Not colour blind	25	30	55
Total	29	36	65

   c No. Fraction of females who are colour blind ($\frac{1}{6}$) is
      bigger than that for males ($\frac{4}{29}$).
7 a 1982 1:1.09   1987 1:1.08
      1992 1:1.07   1997 1:1.06
   b Life expectancy was greater for females than males
      in the years shown.
      Life expectancy increased for both males and
      females between 1982 and 1997.
   c 80.6 (1 d.p.) assuming ratio continues to change
      in the same way.

Reduce to unity to find one part

1 a £12 690   b £21 150
2 a £1500   b £500
3 144
4 a 16   b 2
5 a 65   b 104   c C 3:23

Reduce to unity to scale any quantity

1 £12.80
2 405 seconds
3 £15 000
4 40 minutes
5 95 minutes

6 a 120 g   **b** Sud-U-Like (KtC:114 g/£)
7 a Size 10 would give height 212.5 cm
   b No (this would be nearly 7 feet tall)

Several quantities can be scaled in the same proportion

1 80 spelling, 64 grammatical errors
2 1.6 litres orange, 0.8 litres pineapple, 4.8 oz fruit
3 a 750 g spaghetti, 450 g tomatoes, 360 g beef
   b 45 minutes
4 a i 9045 kJ   **ii** 88.2 g
   b 33 biscuits

2.1.3 Working with percentages

Finding a percentage of an amount may involve more than one step

1 14 members
2 174
3 a 2000   **b** 3000
4 4
5 £71
6 i True since 60% is greater than half (50%)
   ii May not be true. 60% of 20 = 12 people
     Those drinking decaff may all be women.
   iii May not be true – they could drink something else
     e.g. tea
7 a 80 notes   **b** £1680   **c** A £1580
8 The suppliers are not telling the truth.
9

	Gold	Silver	Bronze	Total
UK	1	4	8	13
USA	27	6	7	40
USSR	12	9	6	27
Total	40	19	21	80

An amount can be increased or decreased by a percentage

1 42p
2 Shop 2
3 13 cm by 19.5 cm
4 a £3.52
   b Store's detergent is 48p more expensive.
5 a 4320   **b** 4104
6 950 cd
7 a £24 960   **b** £104
8 a

	Under 16	16–24	25–34	Total (millions)
1997	41%	27%	32%	(28.4)
1998	42%	25%	33%	(27.8)

   b i true   **ii** true   **iii** need more information
9 £5671
10 a 5434
   b 521 Some people have moved.
   c 7.5%

A multiplier can be used for repeated changes

1 £267.65 to nearest pence
2 £6264 to nearest £
3 a i £6262.92   **ii** £7889.48 (nearest pence)
   b 9 years
4 8 bounces
5 a Length = 1.97985, Width = 0.73365 m
   b 3 hours

VAT is $17\frac{1}{2}$%

1 a £28   **b** £188
2 a £76   **b** £89.30
3 a £18   **b** £21.15
4 a £14.70   **b** £248.70

A change can be expressed as a percentage

1 a 20%   **b** 40%   **c** 37.5%
2 30%
3 25%
4 a 25%   **b** Millennium celebrations
5 a 50%   **b** Children in school in September
6 a True   **b** False   **c** True
7 B 75
8 a 7% (nearest %)
   b The change is not an improvement because the time has increased.
9 a Scotland   **b** Northern Ireland   **c** England

2.1.4 Working with large and small numbers

Very large and small numbers are easier to work with in standard form

1 a 500 000   **b** £4 600 000
   c 19 000   **d** 8 000 000 000
   e £3 250 000 000   **f** £2 250 000 000
   g 5 350 012   **h** 450 068 005
2 a 1 600 000   **b** 0.000 001 6
   c 400 000 000   **d** 0.000 000 04
   e 714 000   **f** 0.000 071 4
3 a 6.2×10^7 **b** 9.13×10^4
   c 8×10^8 **d** 8.5×10^{-4}
   e 5×10^{-6} **f** 1.586×10^{-3}
4 a South America   **b** Asia
5 Electron, Proton, Neutron
6 a 8530   **b** 0.0668   **c** 70
   d 0.000 05   **e** 27 000 000   **f** 0.08
7 a 2×10^{-6} **b** 1.87×10^9
   c 1.216×10^{-11} **d** 3×10^9
   e 6.25×10^{-6} **f** 7×10^6
8 a 1.5×10^{11} metres
   b 2.998×10^8 metres per second
   c 500 seconds
9 a 2.4×10^6 million customers
     1.5×10^9 calls per year
   b 1.7 calls per customer per day (1 d.p.)
10 a i 1 296 000   **ii** 812 800
   b 7.3% (1 d.p.)
   c Answer to **b** is between 6.4% and 8.1%

2.1.5 Working with negative numbers

1 **a** −2.5°C   **b** −0.5°C
2 4000 feet
3 **a**

	Starting temperature	Temperature change	Final temperature
A	5	−12	**−7**
B	−4	**+10**	6
C	**−7**	−5	−12

   b C
4 **a** A   **b** £118
   c −£240.50   **d** −£144.25

2.1.6 Working with units

There are many different units of time

1 **a** 300 minutes   **b** 36 months
   c 156 weeks   **d** 11 years
   e 11 days   **f** 13 weeks
   g 3600 seconds   **h** 2880 minutes
2 **a** 1 hour 50 minutes
   b 3 years 6 months
   c 4 days 12 hours
3 **C** $2000 \times 365 \div 7$
4 more than 1 per second
5 No. There is not enough room.

Times can be given using the 12-hour or 24-hour clock

1 **a** 11:15 am   1115
   b 2:30 pm   1430
   c 3:10 am   0310
   d 6:45 am   0645
   e 7:40 pm   1940
   f 11:55 pm   2355
2 **a** 0718   **b** 1621   **c** 2154   **d** 1200
3 **a** 11:20 am   **b** 1:53 pm   **c** 3:47 am   **d** 10:22 pm

To calculate journey times use hours and minutes

1 9:51 am
2 11:02 am
3 **a** Thursday   **b** 19:12
4 8 pm
5 1315 (or 1.15 pm)
6 2 h 37 min
7 **a** 33 minutes   **b** 3 h 35 min
8 2 h 39 min
9 **a** 2 h 24 min   **b** 1921
10 **a** £9.30 per hour   **b** £262.47

Timetables usually use the 24-hour clock

1 **a** 42 minutes   **b** 1 h 12 min   **c** 1 h 19 min
2 **a** 0951   **b** 0946   **c** 0925   **d** 1117

Metric units are decimal measures

1 **a** 3000 g   **b** 250 m*l*
   c 7.5 km   **d** 3.4 *l*
2 **a** 500 000 cm   **b** 3 200 000 g   **c** 150 m*l*
3 **a** 4050 g   **b** 4.05 kg
4 12 people

5 7 turns
6 **a** 4.6 kg   **b** 4600 g
7 **A** 16
8 **a** 6%   **b** 333.3̇ days
NB No 9, 10 and 11 are approximate answers
9 **a** 3 mm   **b** 21 cm   **c** 25 m
10 **a** 5 *l*   **b** 2 kg   **c** 15 g
11 **a** 3.7 cm   **b** 4.3 cm

Imperial units were used in this country before metric units

1 **a** 48 in   **b** 42 lb   **c** 1 st 12 lb
   d 90 fl oz   **e** 4 gallons 4 pints   **f** 560 st
2 8 lb 1 oz
3 28 minutes
4 17

You can convert between metric and imperial units

1 **a** 90 cm   **b** 24.2 lb
   c 15 cm   **d** 21 pints
2 **a** 2 ft   **b** 10 in
   c 20 miles   **d** 10 kg
3 **a** 50 km   **b** 9 ft
   c 38   **d** 15 in
4 1 ton of feathers
5 48 *l*
6 **C** 8 in by 10 in
7 5.4 m
8 **a** 9 whole tiles
   b Area = 100 cm²
   c 76 titles
9 'Poisonous' Pete Potter

Each country has its own currency

1 £12
2 **a** 1020 francs   **b** 200 francs left
3 **a** 52 000 pesetas   **b** £50
4 **a** £147   **b** 441 guilders
5 **C** Belgium
6 **B** $850 - (2175 \div 3.1)$
7 **a** 1620 guilders   **b** 21 000 Belgian francs
8 No
9 **a** £9.25
   b **i** 2650   **ii** 15 756
   c **i** £456   **ii** £1395

2.2 Using Formulae

2.2.1 Formulae and substitution

A formula shows how two or more quantities are related

1 **a** 1 h 40 min   **b** 1 h 28 min
2 **a** 5.6 m   **b** 5 m
3 **a** **i** £79 **ii** £97   **b** £18 per hour
   c £25 call-out fee
4 **B** $A = 25 + 16h$
5 **a** £35   **b** £15
6 **a** Wednesday   **b** **B** £720
7 **a** Rome (86°F)   **b** Rome (23°F)
   c 300°F, 300°C is too hot

8 **D** $m = 3(p - 2)$
9 **a** 25°C
   b 30°C
   c Since 82 lies between 77 and 86, Prina's answer should be between 25 and 30.
   d She put the second bracket in the wrong place.
   $5 \times (82 - 32) \div 9$ giving 27.7°C
10 **a** 25 watts   **b** 28.8 watts
11 £4477.12 (nearest pence)

2.2.2 Compound measures

1 **a** 56 kph   **b** $\frac{3}{4}$ h or 45 minutes
2 $5\frac{1}{2}$ hours
3 276 kg
4 9.6 psi
5 **a** 20 km
   b Measuring from map is not precise.
   c 120 kph
   d **B** 1.9 hours
6

Departure time	Distance	Speed	Journey time	Arrival time
London: **1000**	240 km	120 kph	**2 h**	Norfolk: **1200**
Norfolk: **1300**	240 km	**75 kph**	**3 h 12 min (3.2 h)**	London: 1612

7 **a** 2500 kg/m³   **b** 0.24 m³
8 **a** 1500 feet   **b** 240 seconds = 4 min
9 0.6 litres or 600 m*l*
10 £244.50

Convert components of compound units one at a time

1 110 mph
2 200 kph
3 **a** 7.5 m/s   **b** 0.2 seconds
4 Yes $S = \frac{D}{T} \approx \frac{1.6}{4} = 0.4$ km/min = 24 kph
5 **a** 187 500 miles per second
   b 8 minutes (nearest minute)
6 **a** 480 m/s   **b, c** student's own answers.

2.2.3 Working with equations

1 **a** $n - 9$ **b** $n - 9 = 16, n = 25$
2 $h + 96 = 252, h = 156$
3 **a** $40l = 480, l = 12$ **b** 560 watts
4 **a** $m - 399 = 27, m = 426$ **b** £240
5 **a** $16\frac{1}{2}$ km
   b $x + 12\frac{1}{2} = 16\frac{1}{2}$ $x = 4$ km
   c Via B (Distances $8\frac{1}{5}$ km, $8\frac{3}{10}$ km)
6 **a** $6c = 15.60$ $c = £2.60$ **b** £0.52
7 **a** $8b$ disks
   b $8b = 256$ $b = 32$
   c 96.9% (1 d.p.)
8 **a** $2x$
   b 5p coins worth $5x$ pence,
   10p coins worth $20x$ pence, Adding $25x = 675$
   c $x = 27$ giving 81 coins.
   d **B** 11:19

Equations can involve more than one operation

1 **a** 3 hours at £p per hour is £$3p$.
   Adding £10 gives £28 so $3p + 10 = 28$
   b $p = 6$ £6 per hour
   c £40
2 **a** £$5t$
   b Spends £3, so takes home £$(5t - 3)$
   c $5t - 3 = 27$ $t = 6$ Neil works for 6 hours.
3 **a** $2w + 200$ grams
   b $2w + 200 = 2000$ Density 900 g/litre
4 **a** 1st lap 50 s   2nd lap $50 + t$ (seconds)
   3rd lap $50 + t + t = 50 + 2t$ (seconds)
   b $50 + 5t = 80$ $t = 6$ Times increase by 6 s
   c 390 s
5 **a** £$2p$
   b £$6p$
   c $6p - 2000 = 1450$ $p = 575$
   Initial investment £575
   d 150%
6 $7d + 160 = 1070$ $d = 130$ cm or 1.3 m

Remove brackets by multiplying out

1 $4(l + 2.5) = 850$ $l = 210$

   Length of each side of the blanket = 210 cm
2 **a** No. of mins spent training in each session $= t - 5$
   No. of mins in 6 sessions $= 6(t - 5)$
   4 h = 4 × 60 = 240 min so $6(t - 5) = 240$
   b $t = 45$ Training sessions last 45 mins
   c £112.50
3 **a** Total mass of each can $= m + 80$ grams
   Mass of 6 cans is $6(m + 80)$
   3 kg = 3000 grams so $6(m + 80) = 3000$
   b $m = 420$ 420 g in each can
   c 15 cans
4 **a** £$2(m + 100)$
   b $2(m + 100) - 125 = 1575$
   $m = 750$ Investment was £750
   c 2.1 times bigger

2.2.4 Solving more difficult equations
Collect terms

1 **a** For h hours work, Plumber A charges £$20h$
   Adding the call-out fee gives £$(10 + 20h)$
   Similarly Plumber B charges £$(20 + 16h)$
   Equal charges mean $10 + 20h = 20 + 16h$
   b $h = 2.5$, 2.5 hours
2 **a** Adding 12 years to Mary's age gives $m + 12$
   This equals 3 times Mary's present age i.e. $3m$
   b $m = 6$ Mary is 6 years old.
3 **a** **i** Sue: £$(x + 3)$ **ii** Glyn: £$2(x + 3)$
   b $x + x + 3 + 2(x + 3) = 55, x = 11.5$
   c Dan £11.50   Sue £14.50   Glyn £29
4 **a** Jim takes $(t - 20)$ seconds
   For Jim $D = S \times T = 15(t - 20)$
   For Pete $D = S \times T = 14.4t$
   b Pete takes 500 s, Jim takes 480 s
   c 7.2 km

The inverse of squaring is taking the square root

1 a $\pi r^2 = 78.5$ $r = 5$ cm (nearest cm)
   b 20 place mats
2 9.7 mm (1 d.p.)
3 a Each year the amount is multiplied by $\left(1 + \dfrac{r}{100}\right)$
   b 6.3% (1 d.p.)

2.2.5 Finding two unknowns

1 a Cost of 4 pencils = $4p$ pence
      Total cost in pence = $4p + r$
      This is equal to £1 (100 pence) so $4p + r = 100$
   b $2p + r = 70$
   c Pencil 15 pence, eraser 40 pence.
2 a i $j + t = 111$ ii $j - t = 39$
   b $j = 75, t = 36$ Jacket costs £75, trousers £36
   c £87
3 a Capacity of teapots = Capacity of coffee pots
      gives $5t = 4c$ so $5t - 4c = 0$
      $5t + 4c = 12$
   b $t = 1.2, c = 1.5$
      Teapot holds 1.2 litres, coffee pot 1.5 litres
4 a $16w + 2t = 104.50$
   b $16w + 5t = 129.25$
   c $t = 8.25, w = 5.5$
      basic £5.50 per hour, overtime £8.25 per hour
   d Time and a half   e £154

More difficult equations may need multiplying first

1 a $2b + 3c = 4$
   b $3b + 4c = 5.86$
   c $b = 1.58, c = 0.28$
      Pint of beer costs £1.58, packet of crisps 28p
   d £7.44
2 a The 5 pence coins are worth $5f$ pence and the
      2 pence coins are worth $2t$ pence
      Total is 100 pence so $5f + 2t = 100$
   b $f + t = 26$
   c $f = 16, t = 10$, 16 5p coins and 10 2p coins
   d 4 times
3 a $6d - 4w = 130$
   b Extra in the account = 245 − 130 = £115 so:
      $5d - 3w = 115$
   c $d = 35, w = 20$ Deposits £35, withdrawals £20
   d 12
4 a On Day one £52s was taken for stall seats and
      £29c for circle seats. Total is £1065.
      so $52s + 29c = 1065$
   b $48s + 21c = 885$
   c $s = 11, c = 17$ Stall seats £11, circle seats £17
   d £975
   e Adding gives $100s + 50c = 1950$ gives exactly
      twice the takings on Day 3.
5 a $s - b = -4$
   b $9.5s - 5b = 52$
   c $s = 16, b = 20$
      Salesman buys 20 hairdryers and sells 16.
   d i £82   ii £4.10

2.2.6 Rearranging formulae

1 $C = T - B$
2 a $F = PA$ b $A = \dfrac{F}{P}$
3 a $C = \dfrac{J}{4.2}$ b 310 kC (3 s.f.)
4 a He has divided first instead of subtracting.
   b $u = v - at$
5 a $r = \dfrac{P}{2\pi}$ b $r = 0.95$ m (2 s.f.)

Rearranging a formula may need several steps

1 a $d = \dfrac{C - 20}{15}$ b 6 days
2 c $h = \dfrac{2A}{b}$
3 a $D = \dfrac{60L}{\pi T}$ b 11 cm (nearest cm)   c 95 cm
4 a 18
   b $T = \dfrac{N - 2}{4}$
   c 6 minutes (nearest minute)
   d Bus arrived during this time interval.
5 a $T = \dfrac{MN}{2.2}$ b 300 kg   c 74.6 kg
6 a $M = \dfrac{C - 25D}{0.15}$ b 320 miles   c 8820 miles
7 a $r = \sqrt{\dfrac{V}{\pi d}}$ b 43 m   c 67 m
   d Rate = 12 m/h   e approximately 8:20 am
8 a 14 m/s   b $d = \dfrac{v^2}{19.6}$ c 22.5 metres
9 a $N = \sqrt{D^2 - E^2}$ b 20 km   c East of NE

Rearranging formulae with brackets depends on the position of the required subject

1 a $b = \dfrac{W - 12t}{6}$ or $b = \dfrac{W}{6} - 2t$
   b 35 hours
2 $F = \dfrac{9C + 160}{5}$ or $F = \dfrac{9C}{5} + 32$
3 a £5408
   b $P = \dfrac{I}{\left(1 + \dfrac{R}{100}\right)^2}$ c £7000
4 a $P = \dfrac{A + 2.5C}{2.5}$ or $P = \dfrac{A}{2.5} + C$
   b £609
   c £87
5 a $t = \dfrac{2d}{u + v}$
   b $v = \dfrac{2d - tu}{t}$ or $v = \dfrac{2d}{t} - u$

6 a Multiply by 1000

$7.24(50\,000 - D) = 1000R$

Divide by 7.24

$50\,000 - D = \dfrac{1000R}{7.24}$

Add D

$50\,000 = \dfrac{1000R}{7.24} + D$

$\dfrac{1000R}{7.24} + D = 50\,000$

Subtract $\dfrac{1000R}{7.24}$

$D = 50\,000 - \dfrac{1000R}{7.24}$

b £300   **c** £8563.54

2.3 Shape and Space

2.3.1 Working with plane shapes

1   12 m (nearest m)
2   420 m
3   **a** 504 m   **b** 20 times
4   262 m
5   10.5 m or 11 m (nearest m)
6   1700 m
7   2.4 m
8   2.5 m

Area is measured in square units

1   **a** 1.5 m²   **b** 3 m²   **c** 3.8 m²
2   16 cm² (2 s.f.)
3   $1\frac{7}{8}$ square inches
4   110 m² (2 s.f.)
5   364 m²
6   **a** 72 m²   **b** 32 m² (2 s.f.)   **c** 41 m² (2 s.f.)
7   **a i** 18 000 cm² (2 s.f.)   **ii** 13 000 cm² (2 s.f.)
    b i 1.8 m² (2 s.f.)   **ii** 1.3 m² (2 s.f.)
8   **a** 4 m   **b** 3 m   **c** 2.4 m   **d** 2 m   **e** 1.6 m

A triangle's area depends on its base and height

1   **a** 700 cm²   **b** 825 cm²   **c** 0.45 m²
2   £332
3   2.0 hectares (2 s.f.)
4   2400 cm² (2 s.f.)

A circle is a plane shape

1   400 m (2 s.f.)
2   **a** 62 mm   **b** 1200 mm² (2 s.f.)
3   **a** 200 cm (2 s.f.)   **b** 500 revolutions (approx.)
4   22 cm
5   **a** 121 m (nearest metre)
    b 8000 m² (2 s.f.)
6   **a** 10.9 cm (1 d.p.)   **b** 327 km
7   16 cm (nearest cm)
8   72 years

9   **a** $r = \sqrt{\dfrac{A}{\pi}}$ **b** 30 mm (nearest mm)

2.3.2 Volumes and solid shapes

Volume is measured in cubic units

1   3.6 m³
2   1.1 m³ (2 s.f.)
3   **a** 600 cm³   **b** 13
4   **a** $10 \times 10 \times 10 = 1000$ mm³   **b** 1 000 000 cm²
    c i 800 mm³   **ii** 0.24 m³
5   **a** 30 litres   **b** 20 fish
6   **a** 60 cm³   **b** 7200 cm³   **c** 120
7   **a** 1.7 cm³ (2 s.f.)   **b** 540 cm³   **c** 300
8   **a** 120 cm³ (2 s.f.)   **b** Boxes A and C
9   22 cm (nearest cm)
10   12.5 cm

Prisms have a constant cross section

1   480 cm³ (1 s.f.)
2   144 cm³
3   200 litres (1 s.f.)
4   **a** 5.5 m²   **b** 17 m³ (2 s.f.)
5   **a** 2000 cm³ (1 s.f.)   **b** 6 cups

6   **a i** $h = \dfrac{V}{\pi r^2}$ **ii** $r = \sqrt{\dfrac{V}{\pi h}}$

    b i 80 cm (2 s.f.)   **ii** 40 cm (2 s.f.)

7   3 bags
8   4 mm

2.3.3 Working with scales

A scale can be given as a ratio

1   **a i** 25 m   **ii** 20 m   **iii** 1 m
    b i 90 m   **ii** 500 m²   **iii** 25 m²
2   **a, b**

	a Dimensions	**b** Area
Games Room	10 m by 7 m	70 m²
Function Room	11 m by 5 m	55 m²
Storeroom	6 m by 5 m	30 m²
Cloakroom	5 m by 3.4 m	17 m²
Corridor	5 m by 1.6 m	8 m²

    c $\frac{1}{6}$
3   Scale drawing of classroom
4   **a** 4 m by 3.2 m   **b** 1.7 m
    c 0.75 m   **d** 0.6 m
    e 1.25 m by 0.9 m   **f** 1.1 m by 0.5 m
NB Answers from maps are approximate.
5   **a** 4.3 km   **b** 5.2 km
    c 6.5 km   **d** 9 km
6   **a** 580 m   **b** 2.6 km
    c 150 m   **d** 2.6 km
    e Low Farm
7   **a** 1:200 000   **b** 1:2 500 000
    c 1:50 000   **d** 1:50 000
    e 1:63 360   **f** 1:253 440
8   Scale drawing of student's bedroom

Enlargements

1 **a** 1.25 **b** 7.5 cm
2 **a** 5 **b** 12 cm
3 **a** 16 **b** 120 cm²
4 **a i** 1.2 **ii** 1.44
   b 6 cm **c** 28.8 cm²
5 **a** 1:100 **b** 50 m²
6 **a** Length 15 cm, Width 9 cm, Thickness 3 cm
   b i 120 cm³ **ii** 405 cm³ **c** 3.375
7 125
8 **a i** 1.44 **ii** 1.728
   b 650 cm² (2 s.f.)
   c 4700 cm³ (2 s.f.)
9 **a** 15 cm **b** 1500 cm³ (2 s.f.)
10 **a** 2.5 **b** 1.4 (2 s.f.) **c** 14 cm (2 s.f.)

2.3.4 Working with right-angled triangles

A right-angled triangle has an angle of 90°

1 2

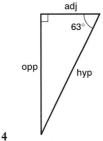

3 4

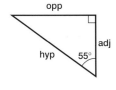

Using angles to find sides

1 7.7 m (1 d.p.)
2 20 m (nearest m)
3 **a** 4.0 m (1 d.p.) **b** 12.7 m (1 d.p.)
4 **a** 75 cm (nearest cm) **b** 77 cm (nearest cm)
5 **a** 5.35 m (2 d.p.) **b** 5.7 m (1 d.p.)

Using sides to find angles

1 11° (nearest °) 2 16° (nearest °)
3 31° (nearest °) 4 80° (nearest °)

Using sides to find sides

1 36 m (nearest m) 2 2.3 m (1 d.p.)
3 31 m (nearest m) 4 1.9 m (1 d.p.)
5 74 m (nearest m) 6 6.6 m (1 d.p.)

3.2 Organising Data

3.2.1 Classifying data

QL = Qualitative, D = Discrete, C = Continuous
1 **a** C (measured) **b** D (counted)
   c D (counted) **d** QL (descriptive)
   e QL (descriptive) **f** C (measured)

2 **a** D **b** C **c** QL **d** D **e** D
3 **a** D **b** C **c** QL **d** C

3.2.2 Frequency tables

1 **a**

Make of phone	Tally	Frequency f
Backch@t	IIII	4
Call Britannia	IIII I	6
FreeCall	III	3
N-Guage	IIII	4
TalkAbout	III	3
		Σf = 20

b Most popular phone = *Call Britannia* (highest frequency)
2 **a** frequencies (in order) 2, 2, 5, 3, 0, 3, 5
   b 1 injection per day
3 **a i** frequencies (in order): 3, 6, 5, 2, 2
   ii frequencies (in order): 7, 6, 4, 1
   b The claim is true.

3.2.3 Grouping data

1 **a**

No. computers in use	Tally	Frequency f
1–3	III	3
4–6	IIII	4
7–9	IIII	5
10–12	IIII	4
		Σf = 16

b the count at 1:30 pm records the highest number of computers being used (12)
2 **a** frequencies (in order): 2, 8, 12, 10, 3
   b 10
3 **a** frequencies (in order): *Busy Buses* 9, 7, 3, 1
           Buses-R-Us: 4, 6, 5, 5
   b *Busy Buses* provides the more reliable service.
4 **a** 2
   b 3 × 11 + 2 × 7 + 15 × 1 = 62
   c No.
   d frequencies (in order): 1, 0, 2, 11

Intervals have class boundaries

1 **a**

Height (cm)	Tally	No. students f
165–168	III	3
169–172	IIII	4
173–176	IIII IIII IIII	14
177–180	IIII	4
181–184	II	2
		Σf = 27

b 173–176

2 a Frequencies (in order): 8, 7, 6, 4
b 16% – there are two times between 30 and 31 seconds
c frequencies (in order): 5, 9, 3, 4, 4
d The second table shows greater efficiency.
3 a 9.95 and 10.15, 10.15 and 10.35, 10.35 and 10.55
b frequencies (in order): 8, 12, 5, 7
c i True **ii** May not be true **iii** True
4 a $\Sigma f = 43$
b frequencies (in order): 2, 30, 9, 2
c The 2nd table indicates there are four people seriously under or overweight but does not indicate *how* serious their condition is. Also, the interval '60–99' is too wide to give any useful information.

3.2.4 Checking your work for errors

1 a $\Sigma f = 49$ (an odd number) Shoes are sold in pairs
b Size $9\frac{1}{2}$ since this is the only odd frequency (frequency should be 8)

3.3 Working with Data

3.3.1 The three averages
Mode

1 a *D D D D D E F F F M M M M M*
b mode = The Moon
2 a 50, 50, 60, 60, 80, 80, 80, 90, 90, 90, 90, 100, 100, 120
b mode = £90
c 114 bonds
3 a frequencies (Mon–Fri): 5, 7, 11, 5, 0
b Friday
c Tuesdays
4 a mode = 0 loaves **b** Wholemeal bread
5 a 2, 2, 2, 3, 3, 4, 4, 5, 5, 5, 5
b ? = 2 calculators

Median

1 a 2, 2, 2, 2, 3, 3, 4 has median of 2 pints, a typical value (i.e. a reasonable representative)
b 9.5 piglets – not a typical value
c 184 cm – a typical value
d £1.09 – not a typical value
2 a A, A, B, B, B, C, C, D, D, E, E, N, U
b median grade = C
c No grade between C and D
3 0.45 seconds
4 a 22 students
b 63.5 marks
c Including 42 and 60 would make the median score 62.5
d 64 marks

Mean

1 $(24 + 18 + 29 + 25 + 19) \div 5 = 23$ km
2 a 12 decorations per hour **b** 360 decorations
3 a Imran **b** Paul
4 a 38.8 kg **b** 26 biscuits

5 595 papers
6 61.5 points
7 4 minutes 25 seconds
8 22.85 years
9 20.5°C

Which average is best?

1 a i 4.5 **ii** 5 **iii** 4.3 robbers
b The mode – it is a whole number and more people agree there were '5' robbers than any other number.
2 a i 1 **ii** 0 **iii** 2 minutes late
b The mode – suggests the student was *never* late!
3 a i 68 **ii** 63 **iii** 90 visitors
b The mean – it takes into account *all* the data.
4 a i 90 **ii** 90 **iii** 132
b median or mode
c The mode
5 a i £8000 **ii** £8000 **iii** £16 389 (nearest £)
b The median and the mode
c The mean
6 a median = 3, mode = 2, mean = 3.3 (1 d.p.) The mean (or possibly the median).
b median = 5, mode = 6, mean = 4.4 (1 d.p.) The mode gives the best impression.
c median = 4, mode = 4, mean = 3. The mean gives the best impression.

3.3.2 The range

1 a range = 0.9 marks (= 5.7 − 4.8)
b range 2400 g (= 2800 − 400)
2 a i 10°C **ii** 13°C
b no information on temperatures between 3 hour intervals given
3 a mean/range: Major roads: 19 mins/11 mins Minor roads: 19.2 mins/3 mins
b The minor roads

3.3.3 Analysing a frequency table

1 a i 3 **ii** 1 (= 16th value)
iii 2 = (0 × 6 + ... + 6 × 1) ÷ 31 **iv** 6
b All reasonably central but mode a bit high, median a bit low and mean not one of the values.
2 a i 12 **ii** 12 **iii** 11.8 (to 1 d.p.) **iv** 5
b Mode and median good, central and one of the values. Mean central but not one of the values.
3 a i 1 **ii** 2 **iii** 2.1 **iv** 6
b Mode and median are all reasonable representatives but mean not one of the values.
4 a i 5 **ii** 5 **iii** 4.6 **iv** 7
b All three are good representatives
5 a i 6 **ii** 4 **iii** 4.05
b mode – the die seems to be biased towards a score of '6'
6 a ? = 6 boxes
b i 0 **ii** 0 **iii** 0.44 cracked eggs per box
c the mean
d 880 cracked eggs

Averages can be estimated from a grouped frequency table

1 a Using midpoints 1, 4, 7, 10:
   mean $\approx (1 \times 25 + 4 \times 4 + 7 \times 0 + 10 \times 1)$
   $\div (25 + 4 + 0 + 1) = 1.7$ errors per paragraph
  b $135 \times 1.7 = 229.5$ i.e. 230 errors
  c No of errors per paragraph $= 1.7$ (using 1st estimate) and $\frac{5}{9} \approx 0.55$ (using 2nd estimate). The second estimate provides evidence of greater accuracy.

2 a mean $\approx £150.74$ (nearest pence)
  b modal class = '£120–' (i.e. £120 – £139.99)
  c new mean $\approx £158.28$ (nearest pence)

3 a i mean $\approx 173\,cm$ ii mean $\approx 175\,cm$
  b 1st table = 1st year, 2nd table = 2nd year

4 a 2nd table gives greater accuracy since it uses narrower intervals.
  b 1st table (mean ≈ 5 cancelled trains per day)

3.4 Presenting Findings

3.4.1 Drawing and analysing bar-charts

1 a *(See Fig. 1 on page 195)*
  b Tea with milk (has the highest frequency)
  c $\Sigma f = 140$ drinks
   i $\frac{28}{140} = 20\%$ ii $\frac{63}{140} = 45\%$ iii $\frac{49}{140} = 35\%$

2 a Break horizontal scale
  b 18 students per course
  c 37 courses   d 6 courses

3 a *See website for full answer*
  b 210 students
  c No. Less than half (= 97) the students asked were unable to park more than once. Ideally the college could increase other transport facilities. *(Many possible sensible answers.)*

4 a i and ii *See website for full answer*
  b The bar-chart drawn in ii – the bars appear to have approximately the same heights.

Bar-charts and histograms can be used to show grouped data

1 a *(See Fig. 2 on page 195)*
  b $16 + 26 = 42$ applicants
  c $\Sigma f = 70$: $\frac{28}{70} = 40\%$

2 a Break horizontal scale   b 35%

3 a *See website for full answer*
  b 350 customers
  c No – the checkout just fails to meet the target.

4 a *See website for full answer*
  b '21–30'   c 37.5%
  d '21–30' may include times between 30 and 30.5 minutes

Comparative and component bar-charts show how two quantities compare

1 *(See Fig. 3 on page 196)*
2 *(See Fig. 4 on page 196)*
3 a i and ii *See website for full answer*
  b A comparative bar-chart
  c Your report should compare the most/least successful countries by type and number of medals.

4 a i Computing   ii Computing
   iii   Health & Social Care
  b Art & Design, Health & Social Care
  c 80%
  d 25%

5 a 1 student
  b 10 distinctions
  c Assignment 1 or 3
  d 96%

6 a 35%   b 10%
  c Similar numbers of males and females passed after one or two attempts. Some females required more tests before passing.

7 a 16 teams
  b Mayfield Utd. win more games at home than away. They also lose less games at home than away.
  c 42 points

A frequency polygon can be drawn on top of a bar-chart or histogram

1 a mid-points are 1.72, 1.77, 1.82, 1.87, 1.92
  b *(See Fig. 5 on page 197)*
2 a 28 students
  b mid-points are 7.5, 13.5, 19.5, 25.5, 31.5, 37.5

Frequency polygons can be used to compare two sets of data

1 a Paris. The polygon for Paris peaks around 22°C–22.5°C
  b 1°C: the polygon for Paris is approximately 1°C further to the right than the polygon for London.

2 a Polygon B
  b 25 students
  c 160 mmHg
  d $179.\dot{9} - 160 = 19\,mmHg$ (working in whole mmHg)

3 a Group A   b Group A
  c Yes: Group A generally finished earlier than Group B. This suggests Group A worked more productively (but no more accurately) than Group B.

3.4.2 Drawing and analysing pie-charts

The angles in a pie-chart represent frequency

1

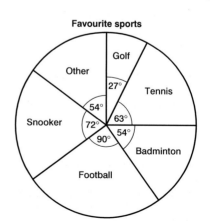

Favourite sports

2 **a** ? = 11 students

    b Angles are: 40°, 88°, 168°, 64°

3 Angles are 60°, 104°, 76°, 91°, 29°

4 **a** ? = 7.5%

    b 90°, 144°, 54°, 45°, 27°

    c *See website for full answer*

5 **a** Angles are: 72°, 54°, 40.5°, 108°, 85.5°

    b Angles are: 79.2°, 43.2°, 28.8°, 122.4°, 86.4°
       {round to whole angles as appropriate}

A pie-chart shows the proportion of each 'part' out of the 'whole'

1 A = poisoning (25% = 90°)

    B = shooting (half the size of 'poison' slice)

    C = strangulation = largest slice

    D = stabbings (remaining slice)

    E = drowning (one-third the size of 'shooting' slice)

2 **a** As $\frac{15}{60} = \frac{10}{40}$, both tables show an equal proportion for '£5'

    b Table 3.65: (use proportion for £20)

3 **a** 144 cars

    b £212.40

4 Use 48° = 360 i.e. 1° = 7.5 papers

    a True – 32° × 7.5 = 240 papers

    b True

    c May be false – some students may sit more than one exam.

    d False – the 'English' slice accounts for less than half the pie-chart. 2700 ÷ 2 = 1350 students

3.4.3 Drawing and analysing line graphs

Drawing a line graph

1 *(See Fig. 6 on page 197)*

2 *See website for full answer*

3 break vertical scale

Analysing a line graph

1 **a** $\Sigma f = 48 \times 1000 = 48\,000$ turkeys

    b $\frac{20}{48} \approx \frac{20}{50} = 40\%$

    c Traditionally, turkeys are more in demand at Christmas and Easter.

2 **a** **i** 1 − −7 = 6°C

       ii 6°C

    b temperatures are only recorded every 30 minutes

    c about 9:45 am

    d The temperature may reach 0°C before 9:45 am.

3 **a** 17°C

    b 9th and 12th of June

    c **i** Yes

       ii No – No information about highest temperatures given.

4 **a** The line indicates the charge increases steadily from point A to point B. In fact, separate charges apply for each time interval.

b Break the vertical scale. Horizontal lines
@ 50 pence from 0 to 1 hours,
@ 70 pence from 1 to 2 hours etc…

5 **a** Males earn on average more than females of the same age.

    b **i** 21 years   **ii** 16 years

    c £50 per week

    d 20 years and 22 years

6 **a** The number of births is decreasing whilst the number of deaths is fairly stable.

    b **i** Yes

       ii No

       iii No – the population may have changed due to emigration and immigration

3.4.4 Drawing and analysing scatter graphs

Drawing a scatter graph

1 *(See Fig. 7 on page 198)*

2 **a** Time (hours): 2.5, 3, 1.5, 2.25, 1, 3.5, 3.75, 2.5, 4.5, 2

    b *See website for full answer*

Analysing a scatter graph

1 *(See Fig. 8, page 198 and Fig. 11 on page 199)*

2 Ornamental pots: The line slopes downwards from left to right because sales will fall with inflated prices. Milkman: the line slopes upwards from left to right because the time taken to complete a round increases with the number of customers.

1 **a** The price increases with the mass

    b 2.3 kg   **c** £1.80

2 **a** Exam results improve with attendance

    b 68%   **c** 50%   **d** 64%

3 **a** 7 workers

    b £3500

    c saves £80

    d A indicates 12 workers would require no days.

4 **a** £70 million

    b approx £90 million (extend the line to reach 30 gold medals)

    c When taken back, the line cuts the 'medal' axis is between 4 and 5. The claim is not valid because of the break in the horizontal axis.

5 Graph 2: the more police there are, the fewer the number of unsolved crimes.

3.5.1 Comparing two sets of data

1 **a** **i** mean score = $\frac{384}{12}$ = 32 marks

       ii range = 47 − 19 = 28 marks

    b The mean score for the second group was higher than for the first group. The second group's scores were less variable (smaller range) than the first group. This indicates the scores for the second group include a number of similar, relatively high scores i.e. the second group was generally more able than the first group.

2 **a** **i** mean speed = 53.5 mph   **ii** range = 20 mph

b Yes. The mean speed and the range of speeds have both decreased.

3 **a** modal class = '4.5–(5.5)'

b The interval '8.5–(9.5)'. Most of the readings fall into the first three intervals.

c Yes. The readings were fairly concentrated around 7 – pH value for clean water.

4 **a** 19 shooting stars (timings start *after* the first sighting)

b mean waiting time = 2.5 minutes

c range = 5.1 minutes

d on average, First night: 1 sighting every 2.5 minutes. Second night, 1 sighting every $\frac{1}{2}$ minute. The sightings on the second night were on average five times more frequent. However, the larger range of waiting times indicates that on the second night a relatively long wait between sightings was occasionally required.

3.5.2 Finding information from bar-charts and histograms

1 **a** mode = 0 cats

b **i** total no. houses = 24 + 23 + 10 + 2 + 0 + 1 = 60 houses

ii total no. cats = 0 × 24 + 1 × 23 + … + 5 × 1 = 54

iii mean no. cats per house = 54 ÷ 60 = 0.9

c median = value in the 30.5th position = 1, as the first two bars represent the first 47 (= 24 + 23) values. The 30th and 31st values are also 1.

d the median (the mode suggests no-one owned a cat when in fact most people owned at least one cat: the mean (0.9 cats) has no real meaning)

2 **a** $\frac{43}{85} \approx 50\%$

b **i** 4 ASs

ii 4 ASs

iii mean = 3.4 ASs

3 **a** 50 cars

b 5.6 cars

c **i** True

ii True

iii False

iv False

4 **a** '2–4' minutes

b 40 patients

c 50%

d 4.5 minutes – estimate because not all at the mid-point

5 **a** $\frac{13}{93} \approx 14\%$

b 2.5 × 6 × 2 + … 17.5 × 1 × 2 = £175 × 1000 + £175 000

6 **a**

Employee	mean	median	mode
Tom	2.75	3	3
Jerry	1.6̇6̇	2	1

b **i** Lowest average and the modes are furthest apart.

ii Medians are closest together (and whole numbers).

c Student's bar-chart.

d Data is bimodal. Mean is 2.2 (i.e. between 2 and 3) – correct. Median is clearly greater than 1 – incorrect. Range is 5 − 0 = 5 – incorrect.

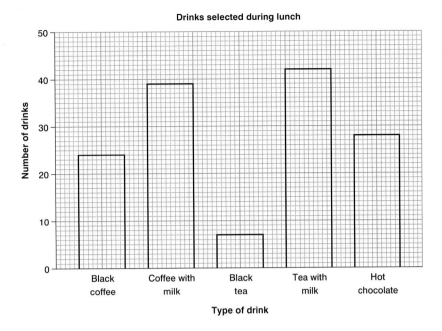

Drinks selected during lunch

Figure 1

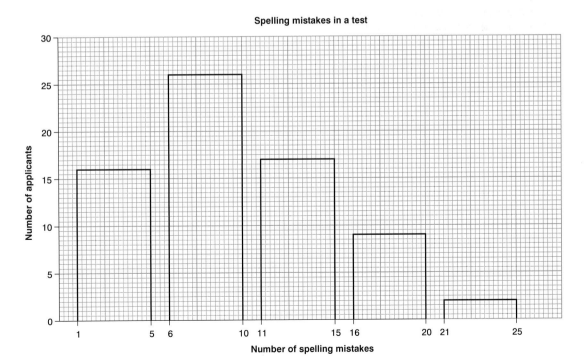

Spelling mistakes in a test

Figure 2

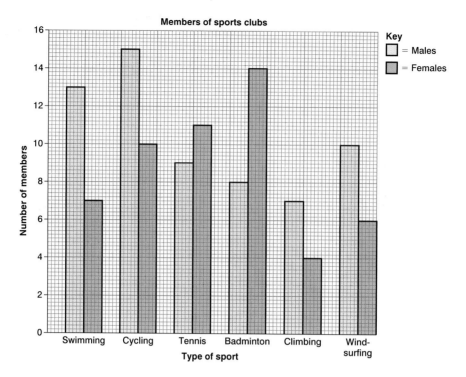

Figure 3

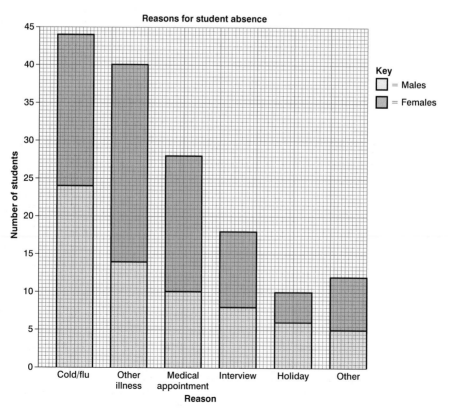

Figure 4

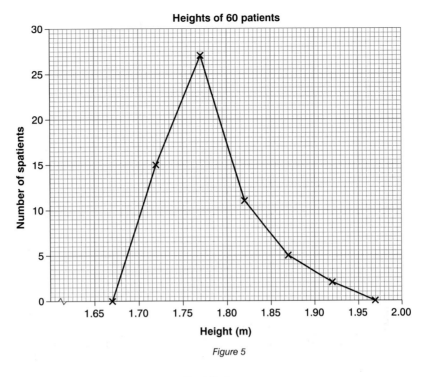

Figure 5

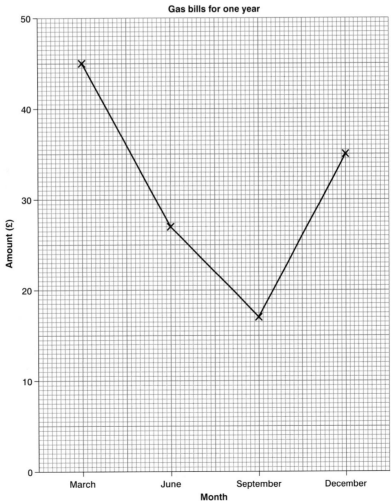

Figure 6

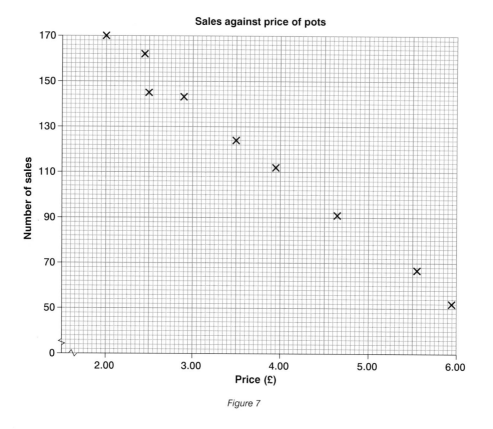

Figure 7

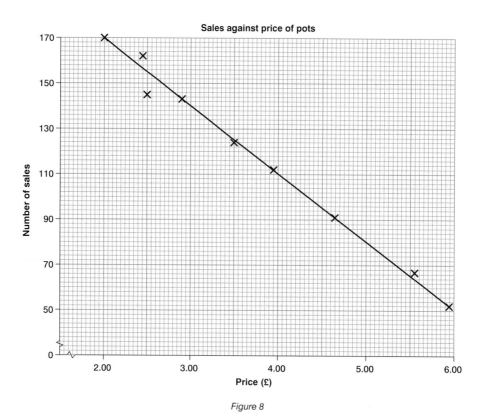

Figure 8

Number of customers against milkround time

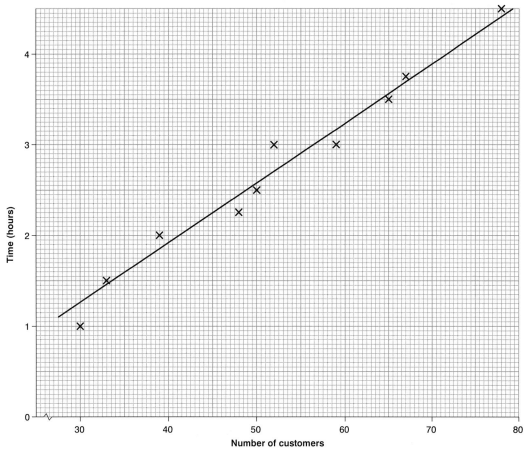

Figure 9

Index